To Leslie ... ...

May 23, 1990

Happy Birthday.

HUMAN CHROMOSOMES
MANUAL OF BASIC TECHNIQUES

Pergamon Titles of Related Interest

Goedde & Agarwal ALCOHOLISM: Biomedical and Genetic Aspects
Kiernan HISTOLOGICAL AND HISTOCHEMICAL METHODS

Related Journals
(Free sample copies available upon request.)

REPRODUCTIVE AND GENETIC ENGINEERING
MOLECULAR ASPECTS OF MEDICINE
CURRENT ADVANCES IN GENETICS AND MICROBIOLOGY
CELLULAR AND MOLECULAR BIOLOGY

HUMAN CHROMOSOMES

MANUAL OF BASIC TECHNIQUES

Ram S. Verma

Chief
Division of Genetics
Long Island College Hospital

Professor
Department of Medicine
SUNY Health Science Center
Brooklyn, NY

and

Arvind Babu

Director
Cytogenetic Laboratory
Division of Medical Genetics
Beth Israel Medical Center

Assistant Professor
Department of Pediatrics
Mount Sinai Medical Center
New York, NY

PERGAMON PRESS

New York • Oxford • Beijing • Frankfurt
São Paulo • Sydney • Tokyo • Toronto

Pergamon Press Offices:

U.S.A.   Pergamon Press, Inc., Maxwell House, Fairview Park,
                    Elmsford, New York 10523, U.S.A.

U.K.   Pergamon Press plc, Headington Hill Hall,
                    Oxford OX3 0BW, England

PEOPLE'S REPUBLIC   Pergamon Press, Qianmen Hotel, Beijing,
OF CHINA   People's Republic of China

FEDERAL REPUBLIC   Pergamon Press GmbH, Hammerweg 6,
OF GERMANY   D-6242 Kronberg, Federal Republic of Germany

BRAZIL   Pergamon Editora Ltda, Rua Eça de Queiros, 346,
                    CEP 04011, São Paulo, Brazil

AUSTRALIA   Pergamon Press (Aust.) Pty., Ltd., P.O. Box 544,
                    Potts Point, NSW 2011, Australia

JAPAN   Pergamon Press, 8th Floor, Matsuoka Central Building,
                    1-7-1 Nishishinjuku, Shinjuku-ku, Tokyo 160, Japan

CANADA   Pergamon Press Canada Ltd., Suite 271, 253 College Street,
                    Toronto, Ontario M5T 1R5, Canada

First printing 1989

Library of Congress Cataloging in Publication Data

Human chromosomes : manual of basic techniques / edited by Ram S.
    Verma, Arvind Babu.
        p.   cm.
    Includes bibliographies and indexes.
    ISBN 0-08-035774-1   ISBN 0-08-036839-5 (pbk.)
    1. Human chromosomes--Examination--Laboratory manuals. 2. Human
chromosome abnormalities--Diagnosis--Laboratory manuals.
3. Cytogenetics--Technique. I. Verma, Ram S. (Ram Sagar) II. Babu
Arvind.
RB44.H85   1989
616′.042--dc19
                                                        88-25517
                                                        CIP

Printed in the United States of America

∞™ The paper used in this publication meets the minimum requirements of
American National Standard for Information Sciences -- Permanence of
Paper for Printed Library Materials, ANSI Z39.48-1984

"Put it before them briefly so they will read it,
clearly so they will appreciate it,
picturesquely so they will remember it
and, above all, accurately
so they will be guided by its light."

JOSEPH PULITZER

Dedication

To our parents

Contents

Preface

Research on human chromosomes has been continuing for a century now. Nevertheless, it is difficult to identify the person who first studied human chromosomes. However, discrepant views by various cytologists concerning the diploid number of human chromosomes continued to cause controversy until resolved in 1956. The presence of an extra G-group chromosome in the Down syndrome began an exciting era with the birth of *Clinical Cytogenetics*. Recent advances in staining techniques have significantly enhanced the biologist's abilities to understand the biologic processes on which survival depends. The identification of individual chromosomes remained an arduous task until the early 1970s. The Q-banding technique opened a new avenue that enhanced the understanding of the structural organization of chromosomes. Chronologic progress for refining these bands through application of various techniques became an intellectual exercise for basic scientists, whereas to clinicians the aberrant chromosomes gained status as the basis of some human disease. The astonishing progress achieved toward our understanding of these diseases at the chromosomal level has stimulated researchers to further explore the human genome at a molecular level, establishing genotype-phenotype relationships.

Our purpose in writing this book was to bring together all the various basic techniques used in research of human chromosomes so that novel developments could be clearly presented. At present, there is no such comprehensive source covering these advances. Moreover, the methods published during the last 15 years are annotated in journals that do not encourage detailed descriptions. We are attempting to fill this void.

We have deliberately restricted the theoretic aspects of each technique, but have rather diligently described the technical aspects in depth. We selected procedures that in our experience proved to be the most reliable, reproducible, and simplest to apply in a routine clinical cytogenetics laboratory. These techniques are drawn from many publications but are in daily use in both of our centers. We have attempted to maintain a balance between the scientific basis of the procedure and its applicability. Each topic has an introduction with key references. We invited distinguished scientists to report on those techniques that we do not practice routinely. Methods are described step by step with appropriate comments.

We begin by describing various tissue culture methods for meiotic as well as mitotic chromosome preparations. "Differential" and "selective staining" techniques are covered separately. Highly specialized techniques that are not used routinely in the laboratory are presented in a separate chapter.

Recently a major breakthrough was achieved in the area of prenatal diagnosis. Chorionic villi sampling during the first trimester of pregnancy, which allows diagnosis to be made earlier than is possible by amniotic fluid sampling, is extensively covered in this book.

Evaluation by banding techniques in individuals whose genetic makeup is in question has certainly become the central theme of a routine cytogenetic laboratory. Nevertheless, chromosomal abnormalities frequently cannot be detected with the available banding technology. One approach in these cases is to use *in situ* hybridization in which DNA sequences can be hybridized to different chromosomes. In addition, various blotting techniques are included for those who are about to venture into the area of molecular genetics. The application of these techniques in clinical medicine is summarized in the last chapter.

This book is primarily written for students and technologists in the field of human cytogenetics who wish to improve their laboratory skills. It will also serve as a reference guide for graduate students who want to pursue this very exciting field. Much knowledge regarding human chromosomes has been gained during the last 15 years, and every possible effort has been made to cover the subject in depth. We hope to share our experience with the readers by including photograph(s) of individual techniques. Nevertheless, we wish to stress that these techniques are associated with several variables. Therefore, it can never be overemphasized that to achieve quality results, one should practice patience, dedication, and, above all, imagination.

Ram S. Verma
Arvind Babu
March 3, 1988

Acknowledgments

Our sincere gratitude to the distinguished scientists for contributing chapters on their expertise. We thank those scientists and publishers who graciously contributed figures or granted permission for figure reproduction. The publisher, scientific editors, and many staff members of Pergamon Press also deserve much credit. Without their involvement this book would not have been complete. My special gratitude to Peter Morrell of Morrell Instruments, Inc., Melville, NY, whose generous contribution has helped to make this project a reality.

Our special appreciation to Roger Dunn for continued support and encouragement during the most trying times of this project; to him we will always be thankful. We owe a debt of gratitude to Michael J. Macera for suggestions and proofreading. Many thanks to our scientific photographer, Jose R. Emmanuelli, for helping to produce the photographs.

The technical assistance of Serpouhi Popescu is gratefully acknowledged. We are thankful to Sonia Williams for repeatedly typing and retyping the manuscript. This project would not have been completed without the understanding and patience of our wives, Shakuntala and Jahnavi. To them we are highly indebted.

Ram S. Verma
Arvind Babu
March 3, 1988

ABBREVIATIONS

AFP	alpha fetoprotein
ASG	acetic-Saline-Giemsa
5-Aza-C	5-azacytidine
5-Aza-dC	5-azadeoxycytidine
5MeC	5-methyl cytidine
ALL	acute lymphoblastic leukemia
ANLL	acute nonlymphoblastic leukemia
AT	adenosine-thymidine base pair
ATP	adenosine triphosphate
bcr	breakpoint cluster region
BrdU	5-bromo-deoxyuridine
BSA	bovine serum albumin
C-bands	centromeric bands
CBG	c-bands by barium hydroxide using Giemsa
cDNA	complementary DNA
CMF-PBS	calcium and magnesium free phosphate buffered saline solution
CML	chronic myelogenous leukemia
CTP	cytosine triphosphate
CVS	chronic vilus sampling
DA	distamycin A
DAPI	4′,6-diamidino-2-phenylindole
DATP	2′-deoxyadenosine 5′-triphosphate
DCTP	2′-deoxycytidine 5′-triphosphate
DM	double minute
DMSO	dimethylsulfoxide
DNA	deoxyribonucleic acid
DNase	deoxyribonuclease
DNTP	deoxyribonucleotide triphosphate
dt	deoxythymidine
DTT	dithiothreitol
DTTP	2′-deoxythymidine 5′-triphosphate
EDTA	disodium ethylene diaminetetraacetate, $2H_2O$
EM	electron microscope
FITC	fluorescine isothiocyanate
FUdR	fluorodeoxyuridine
FudR-5′-P	fluorodeoxyuridine-5′-phosphate
G-bands	Giemsa bands
GTG	G-bands by trypsin using Giemsa
GTP	guanosine triphosphate
HBSS	Hank's balanced salt solution
Hoechst-33258	2′-(4-Hydroxyphenyl)-5-(4-methyl-1-piperazinyl)-2-5′-bi-1H-benzimidazole trihydrochloride
HSR	homogeneously staining regions
IgG	immunoglobulin G
ISCN	An International System for Human Cytogenetics Nomenclature
MEM	Eagles' minimum essential medium
MG	methyl green
MTX	methotrexate
N-bands	nucleolar bands
NOR	neucleolar organizer region

NaAc   sodium acetate
OLB   oligo labelling buffer
PBS   phosphate buffered saline solution
Ph[1]   Philadelphia chromosome
PHA   phytohemagglutinin
PTA   phosphotungstic acid
PVP   polyvenylpyrrholidone
Q-bands   quinacrine bands
QFQ   Q-bands by fluorescence using quinacrine
QFH   Q-bands by fluorescence using Hoechst 33258
R-bands   reverse bands
RBA   R-bands by BrdU using acridine orange
RBC   red blood cells
RBG   R-bands by BrdU using Giemsa
RFA   R-bands by fluorescence using acridine orange
RFLP   restriction fragment length polymorphisms
RHG   R-bands by heating using Giemsa
RNA   ribonucleic acid
RNase   ribonuclease
rNTP   ribonucleotide triphosphate
rRNA   ribosomal ribonucleic acid
SCD   sister chromatid differentiation
SCE   sister chromatid exchange
SCs   synaptonemal complexes
SDS   sodium dodecyl sulfate
SSC   saline sodium citrate solution
SSPE   saline-sodium phosphate-EDTA buffer
TES   tris (Hydroxymethyl)methyl-2-aminoethane
TNE   tris, NaCl & EDTA solution
UTP   uridine triphosphate
UV   ultra violet

Introduction

Research on human chromosomes has been ongoing for over a century (Bostock and Sumner 1980; Taylor 1979). Our purpose is not to document every achievement in chronologic order; numerous interesting and eminently readable publications have already done so (Makino, 1975; Vogel and Motulsky, 1986; Hsu, 1979; Adolph, 1988). Rather, we will highlight the important events that have contributed to our current understanding of human chromosome banding techniques.

Over the last 30 years, human chromosomes have become the primary concern of many biologists and physicians (Hennig, 1987; Gustafson and Appels, 1988). With the emergence of staining procedures in the early 1970s, a unique subspecialty of modern medicine, called clinical cytogenetics, was established. To date, a voluminous literature has accumulated (Yunis, 1977; Hamerton, 1971; DeGrouchy and Turleau, 1984). Clinical cytogeneticists have made profound contributions to the understanding of many aspects of human chromosomes and have gained tremendous recognition during the last three decades (Hirschhorn, 1970–1988; Therman, 1986).

It is difficult to determine who was the first person to study human chromosomes. Some believe it was Arnold (1879), whereas others suggest it was Hanseman (1891). Discrepant views held by various authors in the early part of the century continued to awaken interest among cytogeneticists to resolve the controversy over the diploid number of the human species. Until 1956, the number of human chromosomes was believed to be between 37 and 48. The correct number of chromosomes in a human somatic cell was finally determined by Tjio and Levan (1956), using cells cultured from fetal lungs. In the same year, Ford and Hamerton (1956) examined human meiotic chromosomes and counted 23 bivalents at the first spermatocyte metaphase. Both groups of investigators demonstrated that the correct chromosome number was 46. These advances in human cytogenetics depended largely on technical matters such as the availability of suitable tissue containing an adequate number of cells with the potential for division. In the late 1950s metaphases were obtained from skin fibroblasts. The use of phytohemagglutinin (PHA) to stimulate the peripheral blood lymphocytes to divide (Moorhead et al, 1960) and of hypotonic treatment to obtain better metaphase spreads (Hsu, 1952) were major breakthroughs that permitted a large number of metaphases to be obtained in 69 to 72 hours. The metaphases produced satisfactory chromosomes that could clearly be classified into seven groups, A to G. The X-chromosome was identified within the C-group, whereas the Y-chromosome was placed in the G-group. Thus, identification of individual chromosomes remained an arduous task, because chromosomes within a group resemble one another and only conventional methods of staining with acetoorein were available. Although identification of chromosomes by group was a significant achievement, most of the extra or structurally abnormal chromosomes could not be identified. Several investigators in the early 1960s attempted to identify individual chromosomes by autoradiographic techniques (Lima-de-Faria et al, 1963; German, 1964). However, neither chromosome morphology nor autoradiography provided unequivocal identification for all chromosomes in the human genome. The first advances in human chromosome identification came from Caspersson's laboratory. In a series of papers, he and his collaborators described how chromosomes could be stained by quinacrine mustard (QM), a technique they called Q-banding (Caspersson et al, 1968). In it, chromosomes were differentiated into bright and dark regions, termed "bands." This banding technique encouraged many cytogeneticists to study the mechanisms of banding.

Over a dozen new staining procedures have proliferated since Caspersson's initial discovery (Verma and Dosik, 1982). Thus, the subsequent chapters are exclusively devoted to a detailed discussion of each method. With the application of various staining techniques in individuals with dysmorphic features, several dozen new syndromes have emerged (DeGrouchy and Turleau, 1984). These newer banding techniques have been useful specifically in cases in which structural rearrangements do not alter the size of the chromosomes, in cancer cytogenetics, in clinical medicine, and in prenatal diagnosis using amniocentesis. The cytogenetic basis of disease diagnosis became an essential tool after the discovery of an extra chromosome 21 in Down syndrome. The more recent fascinating area of prenatal diagnosis is to obtain chorionic villi during the first trimester of pregnancy. From this material, chromosome preparations can be made and the status of the fetus assessed rapidly. The chorionic villi sampling procedure is gaining popularity because it can provide a diagnosis earlier than that made by amniotic fluid sampling, drastically reducing the stressful period of anticipating results. In affected fetuses, pregnancy can be terminated earlier while the risk to the mother is still minimal. A detailed description of this procedure is given in Chapter 2.

Molecular biology coupled with cytogenetics has revolutionized the field of molecular genetics (Landegren et al, 1988; Woodhead and Barnhart, 1988). The most spectacular progress has been made in the area of recombinant DNA technology (Child et al, 1988; Perbal, 1984; Davies, 1986; Galton, 1985). The theoretic and practical details of this relatively new approach are covered extensively in Chapter 6.

Banding techniques have become the central theme of every cytogenetic laboratory. However, in many cases chromosomal abnormalities cannot be detected even with all of the available technology. Therefore, in addition to the various blotting techniques, a variety of other strategies are available for identifying these genetic abnormalities. One approach, used extensively since the early 1970s, is in situ hybridization in which DNA sequences can be hybridized to different human chromosomes. This technique plays an important role in the rapid assignment of cloned sequences within the genome and in the characterization of chromosomal abnormalities. Therefore, Chapter 5 will concentrate on the practical uses of this procedure.

It cannot be said that the techniques described in the literature are simple or easy to apply. Journal editors encourage conciseness in the description of new procedures. Consequently, we receive many calls at our laboratories for step-by-step methodology. We therefore felt that an essential need had arisen to produce a compendium of this technical nature.

REFERENCES

Adolph KW (ed): *Chromosomes and Chromatin.* Boca Raton, FL, CRC Press, 1988.

Arnold J: Uber feinere Struktur der Zellen unter normalen und pathologischen Bedingungen. *Virchow's Arch Path Anat* 1879;77:181–206.

Bostock CJ, Sumner AT: *The Eukaryotic Chromosome.* New York, North Holland Publishing Co, 1978, pp 139–171.

Caspersson T, Farber S, Foley GE, Kudynowski J, Modest EJ, Simonsson E, Wagh U, Zech L: Chemical differentiation along metaphase chromosomes. *Exp Cell Res* 1968;49:219–222.

Childs B, Holtzman NA, Kazazian HH, Valle DL (ed): *Molecular Genetics in Medicine.* New York, Elsevier, 1988, pp 3–220.

DeGrouchy J, Turleau C: *Clinical Atlas of Human Chromosomes.* New York, John Wiley, 1984, pp 2–482.

Davies KE (ed). *Human Genetic Diseases: A Practical Approach.* Washington, DC, IRL Press, 1986, pp 1–137.

Ford CE, Hamerton JL: The chromosomes of man. *Nature* 1956;178:1020–1023.

Galton DJ: *Molecular Genetics of Common Metabolic Disease.* New York, John Wiley, 1985, pp 1–137.

German JL: The pattern of DNA synthesis in the chromosomes of human blood cells. *J Cell Biol* 1964; 20:37–55.

Gustafson JP, Appels R (eds): *Chromosome Structure and Function: Impact of New Concepts.* New York, Plenum Press, 1988.

Hamerton JL: *Human Cytogenetics.* New York, Academic Press, 1971.

Hansemann DV: Uber pathologische mitosen. *Virchows Arch Path Anat* 1891;123:356–370.

Hennig W (ed): *Structure and Function of Lukaryotic Chromosomes.* New York, Springer-Verlag, 1987.

Hirschhorn K (ed): *Advances in Human Genetics.* New York, Plenum, 1970–1986, vol 1–15.

Hsu TC: *Human and Mammalian Cytogenetics. An Historical Perspective.* New York, Springer-Verlag, 1979, pp 1–181.

Hsu TC: Mammalian chromosomes in vitro. 1. The karyotype of man. *J Hered* 1952;43:167–172.

Landegren U, Kaiser R, Caskey CT, Hood LC. DNA diagnostics—Molecular techniques and automation. *Science* 1988; 229.

Lima-de-Faria A: *Molecular Evolution and Organization of the Chromosome.* New York, Elsevier, 1983.

Makino S: *Human Chromosome.* Oxford, England, North-Holland, 1975, pp 1–7.

Moorhead PS, Nowell PC, Mellman WJ, Battips DM, Hungerford DA: Chromosome preparations of leukocytes cultured from human peripheral blood. *Exp Cell Res* 1960;20:613–616.

Perbal B: *A Practical Guide to Molecular Cloning.* New York, John Wiley, 1984, pp 487–545.

Taylor JH (ed): *Molecular Genetics: Chromosome Structure.* New York, Academic Press, 1979, pp 1–50.

Therman E: *Human Chromosomes: Structure, Behavior, Effects.* New York, Springer-Verlag, 1986, pp 1–300.

Tjio JH, Levan A: The chromosome number of man. *Hereditas* 1956;42:1–6.

Verma RS, Dosik H: Recent advances in detecting human chromosomal abnormalities by various banding techniques, *in* Sommers SC, Rosen PP (eds): *Pathology Annual*, vol 17, part 2. Norwalk, CT, Appleton-Century-Crofts, 1982, pp 261–286.

Vogel F, Motulsky AG: *Human Genetics: Problems and Approaches*, (ed 3). New York, Springer-Verlag, 1986, p 9.

Woodhead AD, Barnhart BJ (eds): *Biotechnology and the Human Cenome: Innovations and Impact.* New York, Plenum Press, 1988, pp. 1–166.

Yunis JJ (ed): *Chromosomal Syndromes.* New York, Academic Press, 1979.

Tissue Culture Techniques and Chromosome Preparation

With the exception of only a few tissues, most samples used for cytogenetic studies do not have a significant number of endemic mitoses that would enable chromosome preparations without some previous culturing. Thus, tissue culture techniques have become an integral part of chromosome research. The basic concept of tissue culturing has remained unchanged over the years. Nevertheless, technical variations in culture conditions and media have been adopted to (1) obtain a high frequency of cell divisions, (2) achieve better chromosome morphology, and (3) stimulate the cells to divide or enhance the rate of proliferation of cells that are otherwise benign (dormant) in vivo. Highlights of tissue culture methodology are mentioned herein. In-depth information on tissue culture techniques can be found in several specialized books (Freshney, 1983; Barnes et al, 1984; Kruse and Patterson, 1973).

2.1 BASICS OF TISSUE CULTURE

Although cell and tissue culture facilities and techniques could be very elaborate with a number of sterile dust-free compartments, the requirements for routine cytogenetic studies are relatively simple. Nevertheless, it is essential to be aware of basic concepts of sterility that are a part of tissue culture work.

Four primary elements are involved in a tissue culture setup: (1) culture media and media ingredients; (2) culture-ware; (3) work environment; and (4) tissue to be cultured.

Culture media and media ingredients: A number of synthetic media are commercially available that are prepared according to the specifications described by the original authors. These formulated media and other routine media ingredients such as fetal bovine serum, the antibiotics penicillin and streptomycin, and L-glutamine are available in powder form or as ready-to-use solutions. The solutions are supplied in sterile containers and pretested for their sterility and growth-supporting capabilities. For routine clinical, diagnostic, or small research laboratories, it is best to obtain the required media in a diluted form, which eliminates the need for a wide range of sterile facilities, space, and filtration equipment.

The media solutions can be prepared from powder by dissolving the defined amount in triple-glass distilled water and sterilizing by passing it through the cellulose filter with a pore size of $0.2\ \mu$. Filtration is facilitated by generating negative pressure with a vacuum flask connected to a vacuum unit or by applying positive pressure depending on the type of filter equipment. Some basic salt solutions can be sterilized in an autoclave at 15 psi for 15 minutes. However, media containing ingredients other than basal salts should never be subjected to high temperature. Fetal bovine serum can be purchased from commercial suppliers. It has become routine to introduce penicillin and streptomycin at bacteriostatic levels (100 U and 100 μg/ml, respectively) to selectively inhibit the growth of bacteria that may enter the cultures.

Culture-ware: Tissue culture products required for routine cytogenetic work are available as disposable plastics or reusable glassware. Introduction of disposable plastic culture-ware has reduced the laborious cleaning procedures in tissue culture. Plastic-ware is supplied in packages of various sizes sterilized by radiation. It is safe to use if the wrapper around culture-ware is intact; therefore, it is necessary to check the wrapper for damage and perforations before use.

However, some of the glass culture-ware is nondisposable. Therefore, it should be chemically clean (free of chemical and other toxicants) and wrapped in nontoxic paper and sterilized by dry heat at 250°C for at least 2 hours. Rubber corks and plastic and Teflon products should be sterilized in an autoclave at 15 psi for 15 to 20 minutes.

Environment: The culture environment includes the working area (surface and air) and human element. The laboratory may have a separate sterile or dust-free culture cell. Most laboratories use laminar flow hoods. These hoods should be installed where there is the least amount of traffic. They are equipped with an ultraviolet (UV) lamp and filtered air circulation. The surface of the work area is cleaned with alcohol or preferably with a stable antiseptic. The radiation emitted by an UV lamp sterilizes the surface areas, while the air is filtered and forced over the work area to eliminate the dust particles that could enter the cultures. The vertical air-flow type is safer than the horizontal type which blows the air from the work area towards the person working, increasing the human risk of exposure to infections that may be associated with tissue samples. It is important periodically that the lamp be replaced and the filter system be serviced and checked for the rate of air flow, which should always be at the optimum level suggested by the manufacturer. It is necessary to have either a gas burner or a spirit lamp in the laminar hood. The touch-o-matic type of gas burner will lessen the heat load in the work place.

The human element consists of various movements and manipulations required to transfer the solutions and tissue from one culture vessel to another when handling the cultures. It is a good habit to clean hands thoroughly or wear surgical gloves. Wearing gloves provides additional protection when the sample is a potential carrier of infection. Other precautions include flaming the neck and rim regions of the culture flasks and media bottles (taking care not to overheat or melt the plastic-ware) to prevent dust from falling into the container. Contact should be prevented between the tip of the pipette and the surfaces that may not be sterile. The automatic pipetter is handy for transferring fluids. Culture-ware of doubtful sterility should be discarded. Frequent changing of pipettes between different cultures and samples reduces the chance of contamination and cross-contamination.

Tissue sample: Tissue samples should be obtained in sterile condition. Samples such as blood, bone marrow, amniotic fluid, and chorionic villi are relatively easy to obtain by following simple precautions such as cleaning the area with isopropyl alcohol. However, care should be taken with tissues obtained during termination of pregnancy, fetal demise, and certain surgical procedures. If the tissue is suspected of being contaminated, it should be rinsed briefly (about 5 minutes) in rinsing solution consisting of basal salt solution and high-level antibiotics, penicillin, and streptomycin at concentrations of 400 U and 400 μg/ml, respectively. Rinsing time should be kept short, because prolonged exposure of tissue to high-level antibiotics is deleterious to the cells. Subsequently, the tissue should be washed thoroughly in regular balanced salt solution.

2.2 PERIPHERAL BLOOD

Peripheral blood is the most frequently used tissue for postnatal chromosome studies or diagnosis primarily because it is easy to obtain a sample and simple to culture. Normal circulating lymphocytes do not divide under routine culture conditions, but they can be stimulated to proliferate by several lectins such as phytohemagglutinin (PHA), pokeweed mitogen, and concanavalin A. Among them, phytohemagglutinin, an extract of red kidney bean, *Phaseolus vulgaris*, which stimulates the T-cell fraction of lymphocytes, has become the most popular mitogen to study the human chromosomes for diagnostic purposes (Moorhead et al, 1960).

PHA is a glycoprotein that binds to cell membranes and alters the membrane properties. It is noted that the PHA changes the membrane permeability, leading to an increased molecular uptake which in turn triggers the macromolecular synthesis in lymphocytes.

The blood lymphocytes are cultured using either whole blood (microculture) or isolated lymphocytes (macroculture). The choice of method depends mostly on the amount of blood that can be obtained from the subject. Macrocultures obviously involve a large number of nucleated cells and yield more material to prepare the slides. Conversely, it may be difficult to obtain an adequate sample in some cases in which it is necessary to initiate microcultures with a few drops of whole blood. Either way, it is possible to obtain an adequate number of dividing cells to permit cytogenetic diagnosis (also see comments).

2.21   MACROCULTURES

Protocol 2.21.1   Lymphocyte Separation

Procedure

1. Obtain 5 to 10 ml of blood in a heparinized syringe or a green top tube (containing heparin) and mix gently to prevent clotting.
2. Allow to stand at room temperature for 30 to 60 minutes. (The red blood corpuscles [RBC] sediment faster and settle towards the lower half of the syringe or tube, leaving a turbid white lymphocyte-rich plasma at the top.)
3. Collect the top portion of plasma rich with lymphocytes (free of RBC) and make a uniform suspension.
4. Take a small volume of lymphocyte-rich plasma to determine cell density, if needed, and proceed according to the following protocol.

Comments

The blood sample is allowed to stand long enough to separate an adequate amount of lymphocyte-rich plasma. However, if the blood sample is allowed to separate too long, the lymphocytes also sediment, forming a thin white layer or buffy coat at the top of the layer containing RBCs. In these cases, the buffy coat should be obtained along with the supernatant's clear plasma.

Protocol 2.21.2   Culture Technique

Solutions

1. Growth medium

RPMI 1640	100 ml
Fetal bovine serum	25 ml
Penicillin + streptomycin	1.3 ml
(10,000 U/ml and 10 mg/ml, respectively)	
L-glutamine (200 mM or 29.2 mg/ml)	1.3 ml

   RPMI 1640 can be replaced by other media such as MEM with nonessential amino acids, TC 199, McCoy 5A, and RPMI 1630 for suspension cultures. Good mitotic index and chromosome preparations are obtained with any of these media. Growth media can be stored at 2° to 5°C and used for 2 to 3 weeks.

2. Phytohemagglutinin

   Phytohemagglutinin (M-form, lyophilized), as supplied, should be dissolved in the appropriate amount of sterile distilled water suggested by the supplier. It can be stored in lyophilized form at 2° to 5°C for several months. The solution is stored frozen and can be kept for a few weeks if sterility is maintained.

Procedure

1. Prepare the culture tubes by placing 10 ml of growth medium and 0.2 ml of phytohem-agglutinin in each tube. Adjust the medium towards a slightly alkaline pH between 7.4 and 7.6. (Sterile disposable screw cap centrifuge tubes are convenient to handle and eliminate the transfer of culture fluid for harvesting. For routine clinical specimens, initiate triplicate cultures for each sample.)

2. Add 0.3 to 0.5 ml of lymphocyte-rich plasma to each culture tube. The final cell density should be approximately 1×10^6 nucleated cells per milliliter.

3. Mix the contents of each culture tube gently by inverting a few times. Incubate the cultures for 3 days at 37°C in a slanting position. (This position creates more surface area between the liquid and gaseous phases and allows the cells to settle over a larger area of the culture tube, which provides optimal culture conditions for cell growth and proliferation.)

4. Harvest the cultures on the third day (68 to 76 hours from the time of initiation) following protocol 2.21.3.

Comments

Little attention is required for blood cultures, and they can be left undisturbed for the 3-day period. Some laboratories prefer to mix the culture contents once or twice each day. As the incubation proceeds, the pH of the medium decreases as observed by the changing color of phenol red and usually does not require any further adjustment. However, if the cultures become too acidic, loosen the caps of the culture tubes to allow some of the CO_2 to escape till the medium becomes slightly alkaline. If the cultures are to be left unattended, an alternative approach is to start the cultures by inoculating different amounts of cell suspension.

Harvest time: Lymphocyte cultures can be harvested practically at any time beginning 45 hours up to 96 hours from the time the cultures were set. In a general practice, most laboratories use 3-day cultures for cytogenetic studies.

Protocol 2.21.3   Harvesting and Slide Preparation

Solutions

1. Colcemid solution

Colcemid solution (as supplied)	10 μg/ml
or	
Colcemid, lyophilized	10 μg/ml
Diluted with suggested amount of distilled water.	

Solution can be stored at 2° to 5°C for several months.

2. Hypotonic solution (0.56% or 0.075 M KCL)

Potassium chloride	5.6 g
Distilled water	1 L

Store in small quantities. Solution can be stored for a few months if sterilized.

3. Fixative

Methanol—absolute (three parts)	75 ml
Acetic acid—glacial (one part)	25 ml

Prepare fresh before use.

Procedure

1. Add 0.03 ml of colcemid to each culture tube containing 10 ml of medium, mix by gently shaking the tube, and incubate the cultures for an additional 45 to 60 minutes at 37°C.

2. After colcemid treatment, centrifuge the culture tubes at 800 rpm for 8 minutes. (If there are any tough cell aggregates that cannot be dissociated, they should be removed before centrifugation.)

3. Discard the supernatant by pipetting off media, leaving as little medium as possible over the cell button.

4. Resuspend the cell button in 5 ml of hypotonic solution (0.075M KCl) and incubate 10 to 15 minutes in a water bath at 37°C. (Hypotonic solution can be prewarmed to 37°C before use, or the solution stored at room temperature can be used directly to suspend the cells. Either way, it is essential to establish a standard protocol and keep to the same regimen.)

5. Add 5 drops of freshly made fixative to each tube and mix gently by inverting the tubes once or twice. Centrifuge at 800 rpm for 8 minutes. (Cells should be handled very gently following the hypotonic treatment. Any harsh treatment may rupture the cells, leading to many undesirable incomplete metaphases in final preparations. To prevent any such undue damage to the cells, avoid passing the suspension through narrow-tipped droppers or pipettes at this stage.)

6. Discard the supernatant. Disturb the pellet thoroughly by tapping at the bottom of the tube. Resuspend the pellet in 5 ml of fixative (avoid pipetting). Let stand at room temperature for about 10 minutes.

7. Again centrifuge the tubes, discard the supernatant, and suspend the cells in fresh fixative. Repeat this step an additional three times.

8. After the final centrifugation, suspend the cells in a small volume of fixative (approximately 0.5 to 1 ml, depending on the size of the cell button) to give a slightly opaque suspension.

9. Drop 3 to 4 drops evenly on a cold wet slide and allow to dry. After the slide is completely dry, examine under low magnification (10x or 16x) phase objective to check the cell density and spread of metaphase chromosomes. If the cell density is too high, add a few more drops of fixative to the cell suspension. If the cell density is low, centrifuge the suspension and resuspend the pellet in a smaller amount of fixative.

Slide Preparation

It is good practice to examine the first slide or slides before making a large number of final slides. If the preparations are unsatisfactory, there are several variations that may eliminate an existing problem or at least improve the quality of the preparations. Making the slides without preexamination may lead to preparations of poor quality or even to complete loss of important material.

Slide preparation is one of the most important and critical steps in obtaining quality chromosome spreads. Assuming that the cells are subjected to an appropriate length of hypotonic treatment before fixing, ambient temperature and relative humidity play a significant role in chromosome spreading. The following tips can help prepare the slides in a variety of conditions.

1. Cytoplasm surrounding the chromosomes can be due to insufficient hypotonic treatment. This can be dealt with by longer incubation in hypotonic treatment in the remaining samples, if back-up cultures exist, and in future samples. In addition, a few more changes in fixative may improve the spreading to a certain extent. Conversely, if many of the metaphases are incomplete, reduce the exposure of cells to hypotonic solution.

2. In a cold climate, indoor heating leads to very low humidity. The central humidification system may be unable to maintain necessary levels of 50 to 60% relative humidity. The dry ambient weather causes the slides to dry quickly, causing overspreading and many incomplete metaphases. This can be compensated by one or more of the following: (1) The slides may be chilled in a refrigerator before use. (2) Keep a humidifier in the vicinity of the slide

preparation area. (3) After dropping the cell suspension on the slide, gently blow humid air orally.

3. The reverse is usually the problem during hot weather, especially when it is raining. The relative humidity may reach an extent that the slides take extremely long to dry, leading to inadequate spreading of metaphases, persistence of cytoplasmic debris, and the appearance of chromosomes as refractile bodies when observed using phase-contrast microscopy. These problems can be overcome by lowering the ambient temperature (if cooling is efficient) in relation to that of the slides or by using slides kept in slightly warmer (not too hot) water, or by placing the slide on a slide warmer immediately after the cell suspension is dropped on the slide. One may even try to use clean dry slides to drop the cell suspension and reduce the drying time.

Finally, if the cell suspension is too important and insufficient for enough trials, it may be wise to leave the suspension in the refrigerator (2° to 5°C) overnight and try the next morning. If so, the fixative should be changed at least once before attempting to make a slide preparation.

2.22   MICROCULTURES

Protocol 2.22.1   Microculture Technique

Solutions

1. Growth medium
     Same as in protocol 2.21.2.

2. Phytohemagglutinin
     Same as in protocol 2.21.2.

Procedure

1. Prepare culture tubes as described in protocol 2.21.2 (step 1).

2. Obtain a small amount of peripheral blood in a heparinized syringe or green top tube (containing heparin) and place 5 to 10 drops in each culture tube. (The addition of too much whole blood is not suggested, because the excess amount of RBC is detrimental during the harvest and may prevent proper spreading of metaphases.)

3. Incubate the culture tubes in a slanting position at 37°C.

4. Harvest the cultures on the third day. See comments on culture care in protocol 2.21.2.

5. Harvesting and slide preparation methods are the same as those for macrocultures. Follow protocol 2.21.3.

Kinetics of PHA-Stimulated Lymphocytes
Initial incubation of lymphocytes in the presence of PHA for 10 to 15 hours is adequate to stimulate the potential cells to begin macromolecular synthesis and cell division. Removal of PHA from the culture or washing the cells after this period does not decrease or alter the mitotic activity. In routine cultures, the first cell divisions appear at about 40 hours of incubation at 37°C after stimulation. The mitotic index increases up to 2 to 3% and reaches a plateau within about 44 hours. The dividing cells appear over a 6-day period, that is, well beyond 144 hours. However, it has been suggested that the quality of the mitotic chromosomes, with regard to their condensation and morphology, deteriorates rapidly after 120 hours of incubation. Therefore, the optimal time of harvesting the lymphocyte cultures for cytogenetic studies is between 45 and 96 hours. The 70- to 76-hour harvesting protocol is adopted by most laboratories, because there is an increased number of cells ending the second mitosis that yield better chromosomes than do those of the first mitosis during 40 to 50 hours; on the other

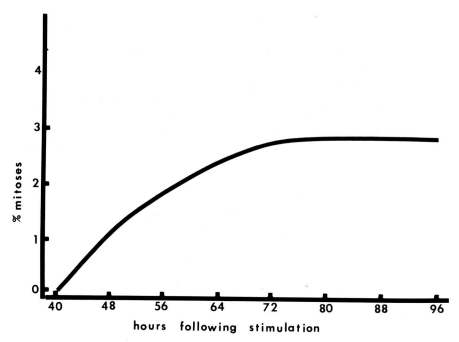

FIGURE 2.2.1. Graphic illustration of mitotic index (% mitoses) of phytohemagglutinin-stimulated lymphocytes at different time intervals (hours following the stimulation of cultures) in human peripheral blood cultures.

hand, the additional incubation up to 96 hours does not provide an added advantage. However, it is optional to follow either schedule to adopt a variation that suits either an experiment or an individual working situation (Figure 2.2.1).

Storage and Transportation of Blood Samples

Peripheral blood samples stored at 2° to 4°C for 4 to 5 days have yielded adequate cell division when stimulated with PHA and incubated at 37°C for 72 hours. During storage at low temperature, care should be taken not to allow freezing of the sample. However, if the sample is available to an investigating laboratory and for some reason the cultures cannot be initiated on the same day, separate the lymphocytes (or whole blood) and store them in the growth medium without adding PHA. PHA can be added when the cultures are ready to be incubated. Samples of peripheral blood for chromosome studies can be transported in foam or another well-insulated package (to prevent exposure to extreme temperature) at ambient temperature within a day or two without deleterious effect.

Chromosome Analysis

At least 20 banded metaphase spreads should be counted and analyzed under the microscope and two metaphases be karyotyped for routine chromosome study. If mosaicism is suspected or noted, a minimum of 30 or more (as needed) cells should be studied.

2.3   BONE MARROW

With the establishment of a relationship between malignancy and chromosomal changes, the cytogenetic studies have become a significant part of a number of hematologic disorders. Cytogenetic studies help in diagnosis, prognosis, therapy, and the like. Bone marrow is the most frequently used tissue.

Although bone marrow samples offer the advantage of immediate harvest due to continuous proliferation, direct chromosome preparations have definite disadvantages. For example, the rate of in vivo cell proliferation is relatively low in most cases, leading to a poor mitotic index. The chromosomes tend to be more condensed, resulting in a short and stubby appearance with diffuse morphology. Furthermore, chromosomes prepared from malignant cells are very sensitive to pretreatment used in various banding techniques. These disadvantages of bone marrow chromosomes have led to variations in culture protocols and synchronization procedures targeted at two particular aspects: to obtain higher mitotic indices with longer chromosomes and to improve the morphology of chromosomes.

Because of the varied sensitivity of leukemic cells to culture conditions and other chemical treatments involved in synchronization procedures, many laboratories adopt more than one protocol to ensure that chromosome analysis does indeed reflect the true cytogenetic status of the disease. Common variations in bone marrow procedures include incubation of the tissue in growth medium for 1 or 2 days, synchronization of the cells, or even stimulation of the cells with one of the B-cell mitogens.

2.31   DIRECT CHROMOSOME PREPARATION

Protocol 2.31.1   Direct Harvest

Solutions

1. Growth medium

RPMI 1640	100 ml
Fetal calf serum	25 ml
Penicillin and streptomycin	1.3 ml
L-glutamine	1.3 ml

2. Fixative

Absolute methanol	75 ml
Glacial acetic acid	25 ml

3. Hypotonic solution

Potassium chloride	5.6 g
Distilled water	1 l

4. Phosphate buffer (0.06 M) (pH 6.8)

Potassium phosphate monobasic (KH_2PO_4)	4.05 g
Sodium phosphate dibasic (Na_2HPO_4)	4.25 g

Instead of sodium phosphate dibasic (Na_2HPO_4), 8.04 g of sodium phosphate dibasic, 7 hydrate ($Na_2HPO_4 \cdot 7H_2O$) or 10.74 g of sodium phosphate dibasic, 12 hydrate ($Na_2HPO_4 \cdot 12H_2O$) can be used to prepare the same amount of buffer.

5. Wright's stain

Wright's stain	0.25 g
Methanol	100 ml

Filter the solution. Store in dark for 1 month at room temperature.

6. Wright's staining solution

Wright's stain	1 part
Phosphate buffer	3 parts

Procedure

1. Obtain 1 to 2 ml of bone marrow in a heparinized syringe.

2. Add 5 to 10 drops of marrow to 10 ml of *unsupplemented* prewarmed medium in a 15-ml centrifuge tube and suspend the cells.

3. Centrifuge the suspension for 8 minutes at 800 rpm and resuspend in *unsupplemented* medium. Repeat centrifugation and finally suspend in 10 ml of *growth medium*. (Washing the cells improves the quality of the chromosome preparation by extracting the cell surface glycoproteins or chalones that may be present on cell membranes.)

4. Expose the cells for 30 minutes at 37°C to 0.05 μg/ml colcemid with the culture tube inclined at an angle.

5. At the end of this period harvest the cultures following the protocol 2.21.3.

6. Prepare the slides following protocol 2.21.3.

7. Stain with Wright's staining solution. Two- to 15-day-old slides are stained for 2 to 3 minutes with one part 0.25% Wright's stain and three parts 0.06 M phosphate buffer at pH 6.8.

8. Chromosomes from leukemic cells are often fuzzy and poorly stained. The sharpness and banding quality can be greatly improved by one or two successive destainings and restainings.

2.32   SHORT-TERM CULTURE

Bone marrow has abundant nucleated cells and fewer red blood cells than does the peripheral blood and usually does not need the isolation of nucleated cells. It is difficult to estimate the cellularity of a marrow sample from physical appearance and without using a microscope. Bone marrow samples from different patients vary in their cellularity to a great extent. This variation makes it difficult to establish a standard volume to set up the cultures. Therefore, we have adopted a simple pretreatment of marrow sample with ammonium chloride to clear red blood cells and to pellet the nucleated cells. This step facilitates estimation of the cell volume and also helps as a washing procedure.

Protocol 2.32.1   Preparation of Marrow Cells

(Following Macera et al, 1988)

Materials

1. Ammonium chloride solution (0.85%)

Ammonium chloride	8.5 g
Distilled water	1 l

Sterilize by membrane filtration. The solution can be stored at room temperature and can be used up to a few months if sterility is maintained.

Procedure

1. Obtain approximately 1 ml of heparinized bone marrow sample in a syringe or a green top tube.

2. Add 10 ml of 0.85 percent ammonium chloride and mix the contents gently. Let it stand for 2 to 3 hours at room temperature or overnight in the refrigerator at 2° to 5°C (never freeze). (Treatment of the sample with ammonium chloride eliminates RBC, and the nucleated cells settle at the bottom.)

3. Centrifuge at 800 rpm for 10 minutes. Discard the supernatant. Wash the cells twice with unsupplemented medium. The cells are ready to be used for culturing as usual.

Protocol 2.32.2   Twenty-Four-Hour Cell Culture

Solutions
Same as in protocol 2.31.1.

Procedure

1. Set up the cultures following steps 1 to 3 of protocol 2.31.1.
2. Incubate the cultures at 37°C for about 24 hours.
3. Treat the cultures with colcemid (0.05 μg/ml) for the final 50 to 60 minutes.
4. Harvest the cultures, prepare the slides, and band the chromosomes following steps 5 to 8 in protocol 2.31.1.

   Slides can be stained by G-banding technique or any other suitable banding procedure.

Chromosome Analysis
In general practice, an average of 20 metaphases are examined thoroughly under a microscope and two karyotypes are prepared for each case. However, the malignant tissues may frequently consist of multiple abnormal clones. The cells with occasional hypodiploid and hyperdiploid chromosome number are carefully analyzed to determine if they represent a random technical artifact or a clone of low population. If the sample consists of multiple clones, further analysis of additional cells is necessary (depending on the availability of the material) to designate the approximate percentage of each clone. At least two cutout karyotypes are prepared for each cell line.

Storage and Transportation
Approximately 1 ml of heparinized bone marrow sample is adequate for routine cytogenetic studies. It is always preferable to work on the sample as soon as possible, particularly if the sample is to be processed for direct chromosome preparation. However, if the sample must be transported to a cytogenetics laboratory, it should be done without further delay following the instructions described previously for peripheral blood samples.

2.4   PRENATAL DIAGNOSIS

Rapid advances in prenatal diagnosis of birth defects have opened a new era in preventive medicine. Clinicians' awareness of genetic and environmental factors presents a promising future for families whose future was uncertain. With the rapid proliferation of newer techniques in prenatal diagnosis, the list of disorders that can be prenatally diagnosed and treated is increasing at an amazing rate. A voluminous literature is available on this exciting subject (Filkins and Russo, 1985; Milunsky, 1988). We will describe herein the technical aspects of culturing amniocytes and chorionic villi.

2.41   AMNIOTIC FLUID

Amniotic fluid is one of the most common tissues used for prenatal diagnosis. Amniocytes present in amniotic fluid, when placed in appropriate culture conditions, attach to the surface and grow in monolayers. The amniocytes have been cultured in T-type flasks for a number of years. However, the need for rapid diagnosis has resulted in a modification of the type of culture vessels and media in which the cells are grown. These variations include (a) growing cell cultures over the cover glass, which enables in situ harvesting, and (b) using culture media that enhance cellular processes like adapting to the culture conditions (dedifferentiation) and growth (proliferation). These approaches make it possible to obtain prenatal diagnostic results from an amniotic fluid sample within 1 week to 10 days that otherwise would have consumed 2 to 3 weeks.

Protocol 2.41.1   Cultures in T-Type Flasks

Growth Media

Culture medium a.

RPMI 1640	100 ml
Fetal bovine serum (FBS)	20 ml
L-glutamine (200 mM)	1.3 ml
Penicillin and streptomycin solution	1.3 ml
(10,000 U/ml and 10,000 μg/ml, respectively)	

Other commercially available tissue culture media such as MEM, Hams F-10, and McCoys can be used instead of RPMI 1640.

Culture medium b.

Chang medium	
Basal medium A	90 ml
Supplement B	10 ml
L-glutamine (200 mM)	1.3 ml
Penicillin and streptomycin solution	1.3 ml
(10,000 U/ml and 10,000 μg/ml, respectively)	

Culture medium c.

Alpha MEM	100 ml
Fetal bovine serum (FBS)	20 ml
L-glutamine (200 mM)	1.3 ml
Penicillin and streptomycin solution	1.3 ml
(10,000 U/ml and 10,000 μg/ml, respectively)	

Culture medium d.

Complete alpha MEM (medium c)	75 ml
Complete Chang medium (medium b)	25 ml

Note: Any of these media can be used to culture amniocytes. Amniocytes grow at an increased rate in modified media. The media are listed in the order of increasing preference by a number of laboratories.

Procedure

1. Place 10 ml of amniotic fluid sample in each prelabeled centrifuge tube. (Usually 20 ml of amniotic fluid is received and distributed into two centrifuge tubes [A & B]. If the sample is smaller, distribute equally into two centrifuge tubes.)

2. Centrifuge the amniotic fluid at 800 rpm for 10 minutes.

3. Note the size of the pellet for the records. Remove the supernatant and save in two new sterile tubes (one tube for analysis of alpha-fetoprotein levels and the other as a backup. The supernatant fluid is stored frozen until use.)

4. Tube A: Add 5 ml of growth medium (either of the above media) and suspend the cells. Transfer the contents to a T25 flask (growth area 25 cm^2), flask A.

   Tube B: Add 7 ml of growth medium and suspend the cells. Distribute the contents into two T25 flasks in volumes of 5 and 2 ml into flasks B and C, respectively. Add an additional 3 ml of growth medium to flask C.

5. Incubate culture flasks A, B, and C in a horizontal position in a CO_2 incubator with caps slightly loose. (The culture flasks should be incubated in two different incubators, flask A in one and flasks B and C in another, to prevent accidental loss of cultures due to instrument failure.)

6. Leave the flasks undisturbed for 4 to 5 days. Following this incubation period, transfer the medium from the flasks along with the floating cells into two sterile centrifuge tubes and replenish culture flasks A, B, and C with fresh growth medium (5 ml per flask).

7. Centrifuge the tubes containing the old medium with floating cells at 800 rpm for 10 minutes. Discard the supernatant and suspend the pellet in 5 ml of growth medium. Transfer the suspension into a fresh T25 flask (flask D).

8. Return all the flasks (A, B, C, and D) to respective CO_2 incubators.

9. Examine culture flasks A and B daily from the seventh day for the number and size of the colonies. Harvest the culture flasks when the number of colonies with active cell proliferation is adequate as judged by examination under an inverted microscope. Renew the medium on the eighth or ninth day in flask C.

Comments

Note: If the colonies are too few and small or show a slow rate of proliferation, renew the medium with fresh growth medium on the eighth or ninth day. It is essential to have healthy and actively proliferating colonies of adequate size for harvesting to obtain adequate cell divisions. If growth is unsatisfactory, the harvesting procedure can be delayed a day or two rather than risk having inadequate chromosome preparations.

It is a general practice not to harvest all the cultures of a particular case on the same day.

Protocol 2.41.2   Harvesting Flask Cultures

Solutions

1. Colcemid solution (10 μg/ml)

   The product is available commercially as lyophilized powder to be dissolved in sterile distilled water or as ready-to-use solution.

2. Trypsin-EDTA solution

Trypsin (1:250)	2.5 g
EDTA	0.38 g
Calcium- and magnesium free-phosphate buffered saline solution (CMF-PBS)	1 l

   Sterilize by filtration and store frozen until use.

3. Hypotonic solution (potassium chloride solution 0.56% or 0.075 M)

Potassium chloride	5.6 g
Distilled water	1 l

   Solution can be used for 2 to 3 weeks if stored at room temperature.
   Note: Other variations in hypotonic solution consist of sodium citrate (0.8 to 1%) or an equal mixture of potassium chloride (0.56%) and sodium citrate (0.1%). Yet other variations, such as half-strength growth medium and the addition of trypsin-EDTA solution to hypotonic solution, are not uncommon. In any event, the duration of hypotonic treatment should be carefully standardized for an existing set of conditions.

4. Fixative
   Same as in protocol 2.21.3.

Procedure

1. Add 0.25 ml of colcemid solution to a culture flask containing 5 ml of medium.
2. Incubate for an additional 2 hours.
3. Remove the medium from the flask and save in a prelabeled centrifuge tube.
4. Add 1 ml of trypsin-EDTA solution (prewarmed to 37°C) to the flask. Wash the cells by tilting the flask from side to side. Remove the solution from the flask and add to the centrifuge tube containing the old medium.
5. Add 1.5 ml of fresh trypsin-EDTA solution to the flask. Bathe the cells with the solution thoroughly by tilting the flask and incubate at 37°C for 5 minutes.
6. Examine the flask. The cells should be free from the surface of the flask. (If not, incubate an additional 2 minutes.) Add the contents of the centrifuge tube to a trypsinized flask, which inhibits the activity of trypsin and prevents further damage to the cells. Mix the contents by pipette to make a uniform suspension. Transfer the suspension back into the centrifuge tube.
7. Centrifuge the tube at 800 rpm for 10 minutes.
8. Discard the supernatant and follow steps 4 to 9 as described in protocol 2.21.3 for peripheral blood lymphocytes.

Comments

Save the medium from the flasks because it may contain floating mitotic cells that are detached from the flask. Alternatively, a small amount of fetal calf serum can be used to inhibit tryptic activity following trypsinization.

Protocol 2.41.3   Cultures on Cover Glass

Solutions

Growth medium as described in protocol 2.41.1.

Procedure

1. Transfer the amniotic fluid sample into two centrifuge tubes and centrifuge at 800 rpm for 10 minutes. Save the supernatant as described in protocol 2.41.1.
2. Suspend the pellet in tubes A and B in 1.5 ml and 2.5 ml of growth medium, respectively.
3. Place 0.5 ml of suspension slowly onto each cover glass in a 35-mm petri dish to prevent any overflow beyond the cover glass. (Prearrange the petri dishes (35-mm diameter) with a 22 mm^2 cover glass. Number the top and bottom halves of the petri dishes before initiating the cultures. Cover glasses are sterilized either by autoclaving in petri dishes between layers of Whatman filter paper (no. 1) or by immersing in absolute alcohol and flaming. If absolute alcohol is used, caution is advised to keep it at a safe distance from the flame.)
4. Incubate the dishes in a CO_2 incubator at 37°C for 24 to 36 hours.
5. Add an additional 1.5 ml of fresh growth medium to each petri dish. Continue incubation.
6. Examine the cultures daily under an inverted microscope after the fifth day from the day of initiation.
7. Harvest the cultures when the growth is adequate.
8. The growth medium in cultures that are not ready for harvest should be renewed on the sixth or seventh day.

Comments

Numbering both halves of each petri dish is important. Sterilization of the cover glasses by autoclaving is preferred to flaming. The cultures are harvested before the colonies grow large and become confluent. Even a slight overgrowth of cultures can significantly retard the spread of chromosomes.

A safe practice in culturing amniocytes on cover glasses involves initiation of one or two flask cultures using 1 ml cell suspension made up to a total volume of 5 ml with growth medium. They serve as a source or backup culture if additional material is required for other studies or if a problem arises with cover glass cultures.

Protocol 2.41.4   Cultures in Slide Chamber

Solutions

1. Growth medium

    As described above.

Procedure

1. Transfer amniotic fluid sample into two centrifuge tubes and centrifuge at 800 rpm for 10 minutes.
2. Save the supernatant fluid as already mentioned. Suspend the pellet in 6 to 8 ml of growth medium. Add 2 ml of suspension to each prelabeled slide chamber. Set up as many cultures as possible.
3. Incubate in a CO_2 incubator with slightly loosened caps.
4. Examine the cultures daily following the sixth day for the number and size of colonies.
5. Harvest the cultures when ready. If not ready, renew the growth medium in each slide chamber with 2 ml of growth medium.

Comments

The slide chambers are designed to grow cells on the surface of the culture vessel, and subsequently the growth area can be used directly for in situ harvesting and microscopic examination. They can be labeled directly and are relatively more convenient to handle than cover glass cultures while providing the advantages of in situ harvesting.

Protocol 2.41.5   HARVESTING IN SITU CULTURES

Solutions
Same as those in protocol 2.41.2.

Procedure

1. Add 0.1 ml of colcemid or velban solution to each 2 ml of culture medium (petri dish or slide chamber).
2. Incubate for an additional 2 hours.
3. Remove the medium as completely as possible by slightly tilting the dish to one side and pipetting off from the bottom area without disturbing the colonies.
4. Gently add 2 ml of hypotonic solution to each culture.
5. Allow to stand at room temperature for 20 to 25 minutes. (Standardization is necessary for a set of laboratory conditions.)

6. Remove the hypotonic solution as completely as possible.

7. Add 2 ml of freshly prepared methanol-acetic acid fixative. Allow to stand for 5 minutes.

8. Discard the old fixative and add 2 ml of fresh fixative.

9. Repeat the change of fixative for an additional two times at intervals of 5–10 minutes.

10. Following the last change of fixative prepare the cover glasses or the slides as follows:

Cover glass: Pipette off the fixative from the petri dish. Lift the cover glass with blunt forceps (facilitated by gently pressing the middle of the bottom half of the petri dish with a finger). Blot off excess fixative by holding the cover glass at an angle to let fixative flow to a corner and gently touching the corner to a blotting paper. Allow to dry under conditions to be described. After drying, mark each cover glass with the case number toward one edge on the back (front contains the cells).

Slide chamber: Remove the fixative from the slide chamber. Lift off the top portion of the flask, retaining the bottom portion containing the cells (used as a slide), and peal off any sealant remaining on the borders of the slide. Blot off excess fixative by holding the slide vertically on blotting paper and allow to dry.

Drying conditions: Drying conditions, which include ambient temperature and relative humidity, are extremely crucial in obtaining proper spreading of chromosomes in in situ harvests. Optimum conditions should be standardized according to the available facilities. A relative humidity of 50 to 60% at an ambient temperature of 24° to 27°C proved to be satisfactory. Further spread of chromosomes is achieved by either blowing air gently over the preparations with a small fan fixed at a distance of at least 1 meter or blowing humidified air onto the cover glass or slide held over a flame at a distance of 8 to 10 cm.

Protocol 2.41.6   Staining the Chromosome Preparations

Solutions
Same as described in protocol 3.12.1.

Procedure

1. Age the preparations by leaving them at 50°C overnight or at room temperature for 3 to 4 days.

2. Stain the preparations following protocol 3.12.1. Cover-glass preparations can be stained without mounting in carriers. Alternatively they can be mounted on a slide with the cell side up before the aging process and treated as regular slides.

3. Preparations are examined as dry slides or after mounting with permount. (The cover-glass preparations stained without mounting should now be mounted on a slide with the cell surface facing either side depending on the type of microscope objective [dry or oil immersion] used for examination.)

Chromosome Analysis and Interpretation
For prenatal diagnosis, a minimum of 20 G-banded metaphases including at least 10 cells from each flask, slide chamber, or cover glass are analyzed for each sample. If preparations consist of at least two or more cells with an abnormal chromosome constitution, they are suspected of mosaicism. Examination of these cases should be supplemented by analysis of cells from additional flasks or from additional colonies in in situ preparations to ascertain the nature of mosaicism.

Sample Transportation
The amniotic fluid samples should be processed for tissue culture as soon as possible. However, it frequently becomes necessary to transport the samples to a cytogenetic laboratory. The samples can be transported in well-insulated or foam boxes if transportation is a mat-

ter of a few hours. If the samples are to be shipped, an overnight mail service with the samples in a foam box that is good enough to protect them from extreme temperatures. During hot weather, wet ice may be included in the package.

2.42   CHORIONIC VILLUS

Note: This section is contributed by Laird Jackson, MD, Longina M. Gibas, and William Coutinho.

2.421   Introduction

Chorionic villus sampling is the removal of a small sample of chorion (placental) tissue for prenatal diagnosis, usually by aspiration biopsy with a thin plastic catheter inserted transvaginally and transcervically into the interior of the pregnant (first-trimester) uterus (Brambati and Oldrini, 1986; Jackson, 1986). The procedure was first attempted in the late 1960s (Mohr, 1968), but its real development began with the demonstration of its increased success using real-time ultrasound to guide the sampling instrument (Hahnemann, 1974; Kazy et al, 1979; 1980; 1982). In 1982, Ward in London (Ward et al, 1983) followed by Brambati in Milan (1987) and Simoni et al (1983) demonstrated the usefulness of the procedure as a clinically applicable tool in prenatal diagnosis of hemoglobin and cytogenetic disorders. Subsequently many others learned and applied the technique in a growing diversity of prenatally detectable fetal conditions. Because the procedure is performed considerably earlier in pregnancy than is the current standard sampling procedure of mid-trimester amniocentesis, it provides a potential patient advantage of decreased anxiety associated with the prenatal diagnosis and a safer pregnancy termination when necessary. Initial evaluations showed that patient morbidity and even discomfort were low, and that serious complications were uncommon (Brambati et al, 1987; Hogge et al, 1986; Jackson et al, 1986; Jackson and Wapner, 1986).

Cytogenetic analysis of the chorionic villi can be accomplished by either the direct technique or the culture method. In the direct technique the Langhans cells of the cytotrophoblast, which actively divide in the first trimester villi, are arrested in mitosis during the short incubation period in vitro in appropriate culture media. Metaphase spreads are then prepared directly from the villi a few hours after sampling. In the culture method the villi are disaggregated by a variety of mechanical and enzymatic methods, and the resulting cell suspension is then used to establish primary cultures. The mesenchymal cells of the villous core released by this procedure are actively proliferative in tissue culture and can be used for cytogenetic analysis 1 to 2 weeks after initiation of cultures. With either method, the sample is first obtained by the obstetrician by one of the standard sampling approaches. The two most common are: (1) transvaginal and transcervical insertion of a soft, thin polyethylene catheter, usually called the transcervical method; and (2) transabdominal insertion of an 18- to 20-gauge needle much as with amniocentesis. Both procedures are performed using continuous, real-time ultrasound guidance and involve insertion of the sampling device into the developing first-trimester placenta, which is called the chorion frondosum. Aspiration by a syringe connected to the sampling catheter or needle draws small pieces of the chorion into the syringe which contains 5.0 ml of nutrient medium (RPMI-1640 with a small amount of heparin). The sample may be transported for at least 24 to 48 hours at ambient temperature in this sterile medium if 5 to 10% fetal calf serum is added.

2.422   Direct Technique

The so-called "direct" technique (metaphase spreads are obtained "directly" from spontaneously mitotic trophoblastic cells or from trophoblasts after only a limited period — hours — of incubation) was originally developed by Simoni et al (1983) from the method of obtaining direct chromosome preparations from fetal membranes of mouse described by Evans and his colleagues (1964). Since then it has undergone several significant modifications. The original method includes a short exposure of the carefully separated and cleaned villi to colcemid, hypotonic treatment, and fixation followed by the release of mitotic cells from the intact villi into the solution by means of brief exposure to 60% acetic acid. The major modifications of this procedure developed in various laboratories involved the following points:

1. Prolongation of the initial incubation time in various types of culture medium from 1 to 2 hours to 24 to 48 hours (Bhatia et al, 1986; Cheung et al, 1985; Karson et al, 1984; Pitmon et al, 1984; Simoni et al, 1984 and 1986; Terzoli et al, 1987).

2. Incorporation of cell synchronization steps into the initial incubation procedure (Gibas et al, 1987; Pergament et al, 1984).

3. Enzymatic treatment of the villi to separate the cytotrophoblast from the intact villus structure (Blakemore et al, 1984).

The method currently used in our laboratory has evolved over a 4-year span as a result of modifications gradually introduced into the original protocol described by Simoni et al (1983). We found that the fluorodioxy uridine (FUdR) synchronization dramatically improves the mitotic index and the morphology of the chromosomes. The protocol, as currently used in our laboratory, is as follows:

1. Transfer the sample to a 35-mm petri dish containing 3 ml of Hanks' balanced salt solution (HBSS) and transfer to the laboratory.

2. Under observation with an inverted microscope or dissecting stereomicroscope, carefully separate the villi from decidua using fine forceps. Wash in HBSS as needed, and divide the sample into two parts (one part for direct preparation and the other for culture).

3. For the direct technique place the villi into a petri dish containing 3 ml of RPMI 1640 medium substituted with 20% inactivated fetal bovine serum and 1% garamycin. The medium is prewarmed to 37°C.

4. Place the specimen in the incubator in an atmosphere of 5% CO_2 and 100% humidity. (All subsequent references to incubator or incubation refer to these conditions.)

5. After at least 5 hours of incubation add 0.1 ml to 10^{-5} M fluorodeoxyuridine (FUdR) solution (final concentration 3.3×10^{-7} M). The working solution is prepared fresh every month and stored in 1.0-aliquots in small plastic vials at -20°C.

6. Incubate overnight for 15 hours.

7. Add 0.1 ml of 10^{-3} M solution of thymidine (final concentration 3.3×10^5 M) and incubate for an additional 5 hours. The working solution of thymidine is prepared and stored in the same way as FUdR.

8. Add colcemid at the final concentration of 0.5 μg/ml and incubate for an additional 2 hours.

9. Remove the dishes from the incubator and gently aspirate the medium using a Pasteur pipette.

10. Slowly add 3 ml of 1% sodium citrate hypotonic solution to every dish. The hypotonic solution should be prewarmed to 37°C. Incubate for 20 minutes.

11. Gently aspirate the hypotonic solution and carefully add 2 ml of fixative (methanol and acetic acid, 3:1) one drop at a time. Aspirate immediately and add a second change of fixative the same way.

12. Place the specimen in a refrigerator at 4°C for 10 minutes.

13. Remove the fixative and leave the petri dish for 1 to 2 minutes at room temperature to allow the remaining fixative to evaporate.

14. Add 0.2 to 0.5 ml of freshly prepared 60% acetic acid. Adjust the amount added depending on the size of the sample. Observe the release of cells from the villi under the inverted microscope and gently agitate the dish.

15. Aspirate the released cells into a 1-ml syringe. Drop the cell suspension on slides prewarmed on a hot plate to 42°C. Spread the suspension over the slide using the bent tip of a Pasteur pipette or a mechanical slide-making device. Typically two to three slides are made from each sample.

16. Incubate the freshly prepared slides overnight in an oven at 65°C.

17. For G-banding, stain with 0.25% Wright's stain solution using a short pretreatment with pancreatin.

This protocol can be adjusted to the work schedule of the laboratory. The incubation times with FUdR and thymidine can vary from 15 to 17 hours and from 5 to 8 hours, respectively, without affecting the metaphase yield. We usually collect the samples from 8 to 12 AM and add FUdR at 5 PM. The thymidine is then added on the following day at 8 AM. The initial incubation before the addition of FUdR should not be shortened below 5 hours; however, it can be prolonged if needed.

2.423 Culture Method

Four to 10 villi are usually received by the tissue culture laboratory after separation from maternal tissue as just described. These villi are placed in 3 ml of Chang's medium (Simoni et al, 1986) in 35-mm dishes and incubated overnight. The next morning the villi are washed in 3 ml of HBSS before the tissue culture process is initiated, as follows:

1. Transfer the villi to a 60-mm petri dish containing 0.0625% trypsin-EDTA in HBSS and incubate for 60 minutes.

2. Gently aspirate and discard the trypsin solution. Add 5 ml of collagenase solution (100 U/ml in MEM) and incubate for 30 minutes.

3. Examination of the villi with an inverted phase microscope should now reveal loose, single cells, grapelike cell clusters and intact villus cores. Pick up the villi with gentle suction against the tip of a Pasteur pipette and transfer to a 25-ml tissue culture flask containing 2 to 5 ml of complete Chang's/RPMI medium (50% each with garamycin and 8% FBS).

4. Gently triturate the sample to obtain a single cell suspension. The flask is observed through the inverted microscope and the cell suspension may be further diluted as indicated.

5. Transfer 0.5-ml aliquots of suspension to 2 to 6 labeled 35-mm plastic petri dishes containing a flame-sterilized coverslip. Return the dishes to the incubator.

6. Add an additional 2.0 ml of medium to each dish after 24 hours.

7. Change to fresh medium in each dish on day 3 after initiation of the culture.

8. On day 5 nestlike colonies of fibroblastic cells are present with a few mitosis. Harvesting of the coverslips is determined by cell growth and mitotic activity as with similar amniotic fluid cell culture (Figure 2.423.1).

9. Add 0.01 μg of colcemid to each dish and incubate for 1 hour.

10. Remove and discard medium from each dish and add 3 ml of pre-warmed (37°C) 0.8% hypotonic sodium citrate solution. Incubate the dishes for 20 minutes.

11. Gently add 2.0 ml of cold, fresh 3:1 methanol acetic acid fixative to the hypotonic solution in each dish and let stand at room temperature for 2 minutes.

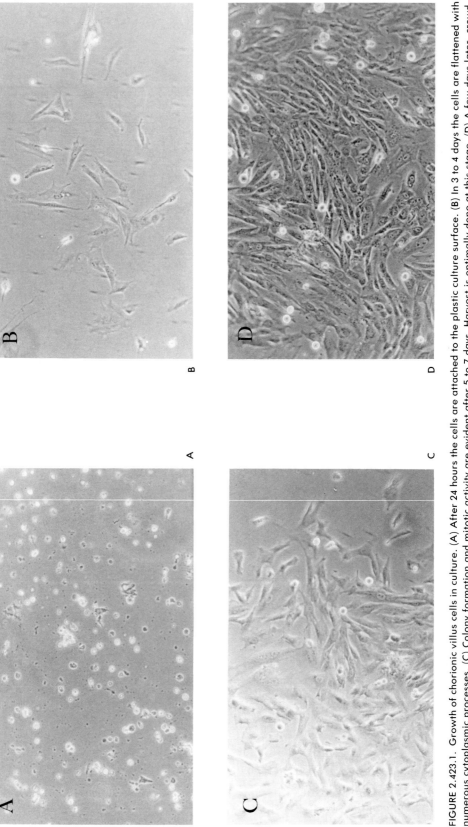

FIGURE 2.423.1. Growth of chorionic villus cells in culture. (A) After 24 hours the cells are attached to the plastic culture surface. (B) In 3 to 4 days the cells are flattened with numerous cytoplasmic processes. (C) Colony formation and mitotic activity are evident after 5 to 7 days. Harvest is optimally done at this stage. (D) A few days later, crowding of cells in the colony will become evident. Despite the presence of numerous mitotic cells in the colony, harvest at this stage will result in suboptimal spreading of the metaphase chromosomes.

12. Aspirate the hypotonic/fixative solution, replace it with 2.0 ml of fresh, cold fixative, and let each dish stand for 20 minutes.

13. Repeat step 12.

14. Repeat step 13 using only 10 minutes of time exposure to fixative.

15. Handling only one dish at a time, aspirate the fixative, invert the dish, tap on absorbent paper to remove excess liquid, and dry the coverslip. The coverslip may be dried by any one of several methods. Some laboratories invert the dish (with coverslip in place) over steam; others blow air directly on the coverslip with or without prior exposure to heat. Exposure to heat is usually accomplished by holding the inverted dish with coverslip above a flame, commonly from an old-fashioned glass chimney oil lamp. Blown air may be adjusted by a small nozzle directing a stream of air from a pump of small capacity. The purpose is to provide relatively slow and controlled, even drying of the coverslip so that the in situ metaphases will spread evenly and well. Trial and error and reasonable control of the room temperature and humidity will provide repeatable results.

16. Remove the coverslip from the dish, label, and incubate on a warming plate overnight at 45°C.

17. Stain with trypsin and Giemsa using a standard protocol.

2.424 Comments

Several technical variables are reported to influence the quality of "direct" chromosome preparations. Bhatia et al (1986) studied the effect of the method of slide preparation on the quantity of metaphases suitable for analysis. They found that the number of analyzable metaphases was highest when the slides were made by allowing the drops of cell suspension to run down an inclined slide (gravity method). Slides made by spreading the cell suspension with a curved glass pipette yielded a smaller total number of metaphases and a higher proportion of disrupted metaphases. These results are contrary to our experience. Although no systematic comparative studies were performed in our laboratory, we did not observe an increased proportion of incomplete metaphases in slides prepared by spreading the cell suspension with the bent end of a glass pipette (rake method). Terzoli et al (1987) described a slide-making instrument consisting of a stainless steel comb and a heated moving stage on which the slides are placed. According to these investigators, the instrument increased the yield of metaphases per slide. We have developed a similar instrument in our laboratory. Gregson and Seabright (1983) developed a modification of the direct technique which includes freezing of the fixed villi at −20°C before treatment with acetic acid and preparations of slides. The specimen is kept frozen for a minimum of 2 hours. Flori et al (1985) used the same procedure. In addition, they rehydrated the specimen before treatment with acetic acid by passing it through a series of alcohol solutions. Metaxotou et al (1987) used overnight storage at 5°C before acetic acid treatment and slide preparation. An interesting variation of the direct technique was described by Blakemore et al (1984). The villi are first digested with trypsin-EDTA and DNase, and the cytotrophoblastic cells are collected for direct analysis. The mesenchymal core elements are used to establish primary cultures. Although this method was originally used only to process single samples, a variation is operating successfully in routine laboratory use (Ledbetter and Ledbetter, personal communication).

The major technical problems associated with direct analysis of chorionic villi are the mitotic index and the quality of metaphase spreads. The problem of mitotic index is easier to overcome than that of the quality of chromosome preparations. Factors that increase the mitotic index, such as prolonging colcemid treatment or increasing its concentration, usually adversely affect the quality of chromosomes by increasing their rate of contraction. Treatment with acetic acid has an adverse effect on the quality of G-banding. Thus, limiting the exposure of chromosomes to acetic acid by speeding up the slide-making procedure (automatic slide makers) will improve the quality of the preparations. Synchronization methods in our experience combine a high mitotic index with good quality of the preparation (Figures 2.424.1 and 2.424.2).

In our initial tissue culture trials, chorionic villi were disrupted by trypsinization only, without prior incubation in Chang's medium. Very few single cells were isolated by trypsin alone, as the cells tend to aggregate in gelatinous clumps. When trypsin and collagenase were used sequentially, many more single cells and cell aggregates were isolated, but these grew slowly and did not proliferate well or demonstrate a good mitotic index. A total of 5 of 100 samples failed to grow, and an average harvest time was 14 days. In contrast, villi that were incubated overnight in Chang's medium and then dissociated by trypsin and collagenase yielded many more single cells and cell aggregates. Initial growth was faster, and the cultures had a high mitotic index. With this method the average harvest time was 9 days, and all samples yielded analyzable results.

This method appears to compare favorably with several of the long-term culture techniques reported in the literature. Niazi et al (1981), using a simple method of maceration of the villi before explantation, indicated a growth period of 5 to 10 days with subculture at an average of 7 days and chromosomal preparation 6 days later. The manual handling required for maceration may be more time-consuming than enzyme dissociation, however. Simoni et al (1983) were the first to report better growth with Chang's medium rather than RPMI. Several laboratories have reported anecdotally that they use variations of the foregoing sequence of enzyme dissociation. Hogge et al (1985) used pronase as the enzyme without subsequent collagenase, at least in their early trials. Their culture times were slightly longer than those reported herein. Schwab et al (1984) reported that their cultures reached confluency on days 8 to 10 and could be used for chromosomal preparation 4 to 8 days after subculture. Mandahl et al (1985) reported that their initiation time was 4 days, subculture 13.9 days, and chromosomal preparation 16.6 days.

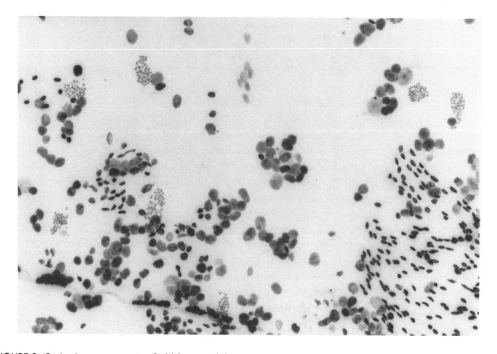

FIGURE 2.424.1. A representative field from a slide prepared by the direct method. Several reasonably well-spread mitoses are evident.

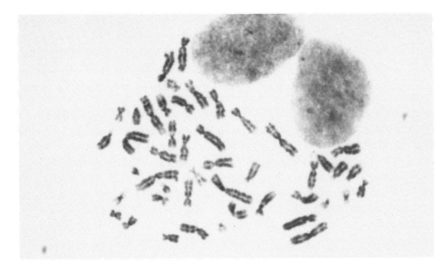

A

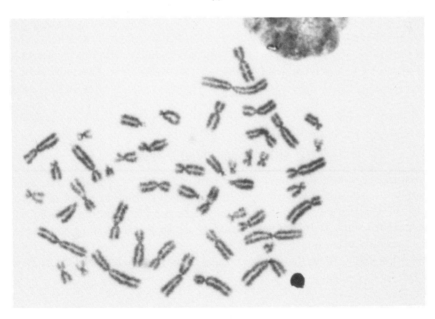

B

FIGURE 2.424.2. Two metaphase spreads obtained by the direct method. (A) Normal female metaphase with condensed chromosomes and good G-banding patterns. (B) A metaphase with 47 chromosomes and a trisomy 18. This spread is more resistant to banding and shows less crispness with a "fuzzy pattern."

With the method described herein, cells do not require subculture or have secondary enzyme dissociation exposure. When sufficient mitoses are present, usually by day 5, sequential harvests may begin. If results are abnormal or a larger quantity of cells is needed, one of the dishes may be subcultured to flasks, providing an early harvest for prenatal diagnosis without the use of enzyme dissociation for subculture.

At the present time, both of these methods are routinely used with every sample processed in our laboratory. The direct method produces metaphase spreads of poorer quality and less reliable mitotic index, but it is not susceptible to significant maternal cell contamination. In fact, the cell cultures have also never demonstrated significant maternal cell contamination in early harvests. Later harvests may be significantly contaminated, which may also occur with amniotic fluid cell cultures. The culturing of only the mesenchymal core is the key to avoiding maternal cell contamination. Removal of this core from the dish containing other floating cells will leave the contaminating maternal cells behind, albeit at the sacrifice of some fetal trophoblastic cells. If appropriate experience is acquired and appropriate preparative care exercised, then culture alone should be the preferred method of chorion villus sampling (CVS) cytogenetics. The reasons for this lay in the biologic frequency with the mosaicism displayed in the primary trophoblastic tissue examined in the direct preparations. Approximately 1% of these preparations will demonstrate some form of mosaicism that will eventually be shown not to exist in the fetus itself (Jackson, 1987). The majority of these mosaicisms will involve tetraploid cell constitutions or unusual trisomies (nos. 3, 9, 12, 22, etc) that should not be compatible with a viable fetus in the second and third trimester. Interpretation is usually fairly straightforward, but further study to confirm the interpretation (amniocentesis or fetal cord blood analysis) will be required. In addition, there will be anxiety for the parents during the study to resolve the interpretation. These instances will be reduced if the culture result is used. There will still be unusual instances of mosaicism, but at a low level. In addition, there have been no instances of abnormal cell lines present in the fetus but absent in cultured CVS analysis, whereas these occurrences have been reported with the direct preparations. In at least one anecdotal report in which only direct studies were used, a liveborn trisomic child resulted. This low rate of diagnostic error is not unexpected in the development of a new technique and is no different from that observed in the development of amniotic fluid cell cultures for prenatal diagnostic use (USNICHD, 1976), and the rate should decrease even more with continued development.

2.5   SKIN AND OTHER TISSUES

Skin and tissues are obtained from adults or fetuses for a variety of reasons. Adult tissues are commonly studied clinically because of an established chromosomal mosaicism in peripheral blood lymphocytes to evaluate for possible tissue-specific variation. In addition, tissues are obtained from fetuses after fetal demise or miscarriage to examine the constitutional chromosomal abnormalities.

Cultures are initiated from these tissues using either dissociated cells after trypsinization or tiny tissue pieces as explants.

Protocol 2.5.1   Cultures Using Tissue Explants

Solutions

1. Hanks' phosphate buffered solution (Hanks' PBS)

Calcium chloride ($CaCl_2$)	0.14 g
Potassium chloride (KCl)	0.40 g
Potassium hydrogen phosphate monobasic (KH_2PO_4)	0.06 g
Magnesium chloride, 6 hydrate ($MgCl_2 \cdot 6H_2O$)	0.10 g
Magnesium sulfate, 7 hydrate ($MgSO_4 \cdot 7H_2O$)	0.10 g
Sodium chloride (NaCl)	8.00 g
Sodium bicarbonate ($NaHCO_3$)	0.35 g
Sodium hydrogen phosphate dibasic, 7 hydrate ($Na_2HPO_4 \cdot 7H_2O$)	0.09 g
Glucose	1.00 g
Distilled water	1.00 l

Prepare the solution and sterilize by filtration or purchase a prepared solution.

2. Growth medium

Same as that in protocol 2.21.2.

Procedure

1. Wash the tissue sample thoroughly in Hanks' PBS to remove toxicants or contaminating blood.

2. Cut the tissue sample into small pieces (approximately 1 mm) with sterile scalpel blades without tearing the tissue.

3. Using the pipette, transfer six to eight pieces of tissue with a minimum amount of fluid into each T25 flask. (Prepare three to four flasks if an adequate amount of tissue is available. If tissue is insufficient, start with a single flask.) Remove excess solution from the flask by positioning it vertically.

4. Distribute the tissue pieces (explants) evenly on the bottom surface (growth area) of the flask and stand the flask in a vertical position to semidry the explants for about 10 to 15 minutes. (Do not allow the tissue to dry out completely. This step is intended to retain the explant adherent to the surface of the flask.)

5. Gently add 5 ml of growth medium to each flask and place the flask(s) in a horizontal position, taking care not to dislodge the explants from the surface.

6. Incubate the flask(s) in a CO_2 incubator. Do not disturb for at least 2 to 3 days.

7. Examine the flasks for cellular growth around the explants and renew the medium. (Cell growth varies considerably among samples. Some tissues may take as long as 1 week before cell migration can be observed.) The explants that fail to remain attached to the flask are either left in the same flask or, if there is adequate growth, are transferred to a new flask(s) by repeating steps 2 to 5.

8. Culture flasks are fed by discarding the old medium and replenishing it with fresh growth medium until cultures reach half-confluency.

9. Cultures are used while in log phase (actively proliferating) either for chromosome preparations or for subculture for further propagation and storage.

Protocol 2.5.2   Cultures Using Dissociated Cells

Solutions

1. Calcium- and magnesium-free phosphate-buffered salt solution (CMF-PBS)
   Same as Hanks' buffer except for the omission of calcium and magnesium salts.

2. Trypsin-EDTA solution

Trypsin (1:250)	2.50 g
EDTA.4Na	0.38 g
CMF-PBS	1 l

3. Growth medium
   Same as in protocol 2.21.2.

Procedure

1. Wash the tissue sample thoroughly in CMS-PBS, place in a fresh sterile petri dish, and cut into small pieces using a scalpel or scissors.

2. Transfer tissue pieces into a trypsinizing flask containing a Teflon-coated magnetic bar and add 15 to 20 ml of trypsin-EDTA solution.

3. Stir the contents of the trypsinizing flask with a magnetic stirrer for 15 to 20 minutes. (Do not use high speed that would cause frothing or injury to the cells.)

4. Transfer the dissociated cells suspended in solution to two centrifuge tubes and add 1 ml of fetal bovine serum to inhibit the trypsin and prevent further damage to the cells. (If tissue pieces still remain in the trypsinizing flask, add 15 to 20 ml of fresh trypsin-EDTA solution and repeat steps 3 and 4.)

5. Centrifuge the tubes at 800 rpm for 10 minutes. Discard the supernatant and suspend the cells in 5 to 10 ml of growth medium, depending on the size of the pellet, to give approximately 1×10^6 cells per milliliter.

6. Introduce 5 ml of cell suspension into each T25 flask and incubate at 37°C in a CO_2 incubator.

7. Feed the cultures on the third or fourth day and once every 2 days thereafter until the cultures reach half-confluency.

8. Cultures are fed 24 hours before use for either chromosome preparations or further propagation.

Chromosome Preparation
Follow harvesting and slide preparation protocols 2.42.2 and 2.21.3 described for amniotic fluid cell cultures and peripheral blood cultures, respectively.

Storage and Transportation
Tissue samples should be placed in tissue culture medium and sent to the Cytogenetic Laboratory. If the sample is to be transported a distance, the container should be completely filled with medium to prevent drying during transportation.

2.6   SOLID TUMORS

Tumor tissues require special handling and processing techniques to obtain satisfactory cell cultures for cytogenetic investigation. Recent advances in cell dissociation protocols and the introduction of several media supplements have increased the rate of success in tumor cell cultures (Leibovitz et al, 1983; McBain et al, 1984; Limon et al, 1986; Gibas et al, 1984).

Protocol 2.6.1   Tumor Cell Cultures

(Following Gibas et al, 1984)

Solutions

1. Culture medium

RPMI 1640	100 ml
Fetal bovine serum (FBS)	20 ml
L-glutamine (200 mM)	1.3 ml
Penicillin and streptomycin solution	1.3 ml
(10,000 U/ml and 10,000 μg/ml, respectively)	

2. Collagenase solution (0.8%)

Collagenase II	120 mg
DNase I	0.3 mg
Culture medium	15 ml

The solution is sterilized by filtration and stored frozen in small aliquots. The solution can be used up to 3 months. The addition of DNase I is optional.

3. Insulin solution (5 μg/ml)

Insulin	5 mg
Culture medium	10 ml

Sterilize by filtration and store at 2 to 5°C *or* prepare the above solution using PBS instead of culture medium and store frozen after filter sterilization. Add 1 ml of insulin solution (5 μg/ml) to 100 ml of medium.

4. Glutathione solution (10 μg/ml)

Glutathione	10 mg
Culture medium	10 ml

Sterilize by filtration and store at 2 to 5°C (do not freeze) *or* prepare the above solution using PBS instead of culture medium and store frozen after filter sterilization. Add 1 ml of glutathione solution (10 μg/ml) to 100 ml of medium.

Procedure

1. Obtain tumor tissue aseptically in a small amount (5 to 10 ml) of sterile saline solution or phosphate-buffered saline (PBS) in a suitable container to transfer the sample to the laboratory. If the tissue sample is too large, trim into smaller pieces before placing them in a sterile solution for transportation. On receiving the sample, if contamination is suspected, the tissue may be rinsed in sterilizing solution (culture medium + high antibiotics) for 2 to 5 minutes.

2. Place tissue sample in a petri dish without medium and trim and discard the fat, normal tissue, and necrotic regions. Mince approximately 0.5 g of tumor tissue using either sharp scissors or scalpel into small pieces of approximately 2 to 3 mm. During the procedure, keep the tissue moist and avoid tearing.

3. Overlay the tissue pieces with an adequate amount (2 to 3 ml) of collagenase solution (0.8% collagenase in culture medium) and incubate at 37°C in a 5 percent CO_2 incubator for 2 to 3 hours. (Some tissues may require prolonged incubation and can be left overnight.)

4. Transfer the entire contents of the petri dish to a centrifuge tube containing culture medium (5 to 10 ml).

5. Triturate the specimen to disaggregate the tissue by passing through a pipette several times. (It is not essential to make the suspension uniform.)

6. Centrifuge the sample at 800 rpm for 10 minutes.

7. Discard the supernatant and suspend the cells and small tissue pieces in 5 to 10 ml of culture medium (approximate final cell density of 2.5 to 3×10^6 cells/ml) containing insulin and glutathione at final concentrations of 5 and 10 μg/ml, respectively.

8. Distribute the suspension in two or three culture flasks depending on the amount of the sample and incubate at 37°C for 1 day.

9. On the following day, discard the medium along with unattached material and replenish with fresh growth medium.

10. Renew the growth medium in culture flasks on alternate days or following a three times a week schedule.

11. Examine the cultures daily after the fourth day for growth and the rate of proliferation using an inverted microscope.

12. Harvest when the colonies are big enough and are actively proliferating. Cultures are usually ready within 1 week to 10 days.

Harvesting

Harvesting is performed by exposing the culture to colcemid (0.01 μg/ml) for 15 hours (overnight) to accumulate enough mitosis. However, in many cases, prolonged exposure to colcemid leads to very short chromosomes. Harvesting and slide preparation procedures are essentially the same as those described for amniotic fluid cell cultures in protocol 2.42.2. An alternative harvesting procedure using synchronization of cultures with methotrexate treatment may be adopted.

Comments

It is important to transport the tumor tissue to the laboratory and initiate the cultures as soon as possible to obtain greater success. Many tumor tissues have necrotic regions that release deleterious toxins. Removal of these necrotic regions before transportation reduces the risk of damage to the remaining tissue. The addition of fungazone to eliminate fungal contamination and a number of other supplements such as polyvinylpyrollidone (PVP) and methyl cellulose (methocel), which offer increased protection to the tumor cells in the transportation medium, have been suggested (Leibovitz, 1986).

Collagenase is preferred over trypsin to disaggregate the cells, because trypsin tends to alter cell surface whereas collagenase destroys the connective tissue without damaging the cells. Mechanical stirring of tissue pieces in enzyme solution can be used to dissociate the tissue. The optimal duration required for collagenase digestion depends on the type of tumor tissue. In usual cases, 2 to 3 hours of digestion is adequate for epithelial tumors, whereas tough tissues such as sarcomas may require prolonged treatment of up to 16 hours. This is best achieved by leaving the minced tissue in collagenase solution overnight. During cell suspension, small tissue pieces may remain and cannot be dissociated. However, these cell aggregates should not be discarded, because several colonies grow from the tissue pieces and cell clumps. Other dissociating media can be used to replace the collagenase solution (Leibovitz et al, 1983).

Chemical supplements, such as insulin and glutathione, added to the culture medium help to improve the growth of tumor cells, but these supplements are not essential. The tumor cells must have enough space for growth on the surface of the flask. Lack of adequate surface area may lead to an excess of stromal fibroblasts which proliferate more aggressively and may overgrow the malignant cells. This can be prevented to some extent by the addition of collagenase to the culture medium or selective removal by careful trypsinization. Alternatively, after having established sufficient cells in the culture vessel, L-15 medium may be used instead of RPMI 1640 in culture medium. L-15 medium has no carbonate-bicarbonate buffer system and therefore, reduces the proliferation of fibroblasts and favors tumor cell growth.

Chromosome Analysis

Chromosome preparations from tumor cultures often may contain normal mitosis of stromal fibroblasts. However, these can be omitted from chromosome analysis during microscopic examination.

2.7 MEIOTIC TECHNIQUE

Note: This section is contributed by M. R. Guichaoua, M. Devictor, C. Toga-Piquet, J. M. Luciani, and A. Stahl.

Current available techniques make it possible to observe almost all stages of meiosis, especially pachytene, diakinesis, and metaphase I as well as metaphase II. Because of the importance of meiosis, mastery of these techniques is desirable, and progress must be made to define pairing modalities and segregation of homologous chromosomes. In a meiotic study, spermatocytes are easier to examine than are oocytes.

Indeed, studies in women are much more difficult to carry out because of the internal position of the ovaries as well as the absence of oogenesis continuity. In fact, most stages in the first meiotic division take place during fetal life. All of the oocytes have reached the diplotene stage several weeks before the end of gestation, and they enter into a long resting phase which lasts a minimum of 10 to 12 years and a maximum of 50 to 55 years. The first meiotic division reaches completion at the moment of ovulation when the ovarian follicle becomes mature. The second meiotic division ends after fertilization. This difference in the chronology and development of oogenesis when compared with the much shorter process of spermatogenesis makes it possible to understand that meiotic techniques are not strictly identical in men and women.

2.71 INVESTIGATION OF METAPHASES I AND II

2.711 Chromosomes from Testicular Material

Male meiosis can be studied in isolated cells originating from fragments of testicular biopsy either directly or after culture. These cells, after being submitted to the action of a hypotonic agent, must be fixed and stained with reagents adapted to DNA.

2.7111 Direct Method
The existence of various modalities in obtaining isolated cells makes it possible to divide all of the technical processes into squashing techniques and cell suspension techniques with air-drying method.

2.71111 Squashing Technique. These techniques were the first to be used. As early as 1952, Makino and Nishimura were able to observe satisfactorily meiotic chromosomes after pretreatment with water.

Progress in squashing techniques subsequently permitted Ford and Hamerton in 1956 to reveal 23 bivalents in spermatocyte I. They therefore confirmed the observations of Tjio and Levan who the same year established the human diploid number at 46. Although these techniques were later used with success, they were abandoned for air-drying methods.

2.71112 Cell Suspension Technique with Air Drying. This technique was initially described by Evans et al in 1964 and modified by Ferguson-Smith (1964), Hulten et al (1966), and Ford and Evans (1969). With all these techniques, a cell suspension could be obtained and successively submitted to hypotonic solution, fixative, and DNA staining.

We made several modifications in the Ford and Evans technique (1969), the most important of which is to use potassium chloride as a hypotonic agent (Luciani et al, 1971). This method gives excellent results at all stages of meiosis, especially metaphases I and II (Figures 2.71112.1 and 2.71112.2).

- Testicular fragments with a volume of 20 to 30 mm^3 are immediately immersed in a hypotonic KCl solution of 0.44% at room temperature and sent to the laboratory as soon as possible. The duration of KCl treatment varies between 10 and 30 minutes. This time can, without inconvenience, be stretched to 3 or 4 hours, making it possible to do longer-range meiotic examinations.

- The seminiferous tubules are teased with forceps or scissors. The suspension is transferred to a conical centrifuge tube. It is best to eliminate the large remaining tubular fragments by sedimentation for 1 or 2 minutes. The supernatant is transferred to another conical centrifuge tube.

- The cell suspension is centrifuged at 1,000 rpm for 7 minutes. The supernatant is discarded, and the pellet is carefully suspended in a small volume of supernatant residue.

- The cells are then fixed in 1 ml of methanol and glacial acetic acid (2:1 volume). When the cell suspension is homogeneous following pipetting, 5 to 6 ml of fixative is added.

- The suspension is then centrifuged at 1,000 rpm for 7 minutes. The supernatant is discarded and the pellet is again placed in suspension in a small residual volume of supernatant.

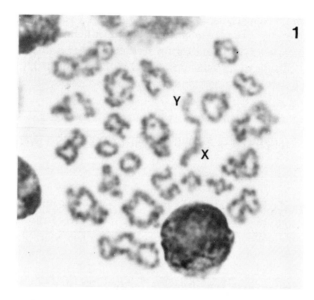

FIGURE 2.71112.1. Metaphase I from a human spermatocyte, showing 23 bivalents (magnification ×1500).

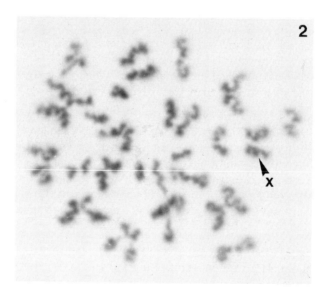

FIGURE 2.71112.2. Metaphase II from a human spermatocyte showing 23 chromosomes (23,X) (magnification ×1500).

- Spreading can be carried out from the final suspension according to the air-drying technique on precooled slides. Phase contrast can be used to check the quality of spreading and cell concentration and then, if necessary, to further dilute the suspension. The air-dried slides can be stored or stained immediately; either Giemsa solution, lactoacetic orcein, or toluidine blue can be used. The speed of this technique as well as the quality of the preparations make it a method of choice for systematic meiotic investigations. Moreover, it should be mentioned that slides prepared in this way are perfectly suitable for fluorescence techniques and cytochemical analysis.

Caspersson et al (1971), using quinacrine mustard staining, demonstrated that it was possible to identify all of the bivalents in the meiosis of man. This same technique permitted Pearson and Bobrow (1970) to end the discussion on the way X and Y chromosome arms are associated at diakinesis—metaphase I by showing that the relationship was established by their short arms.

Centromeric heterochromatin visualization can be obtained using the Arrighi and Hsu technique (1971). Combination with the previous fluorescence method led to specific identification of each bivalent at metaphase I (Gagné et al, 1973; Hulten, 1974).

Among cell suspension techniques, that of Meredith (1969) deserves attention, because this original method has the advantage of suppressing the number of centrifugations required by the technique of Evans et al (1964) or those that it inspired.

Meredith's method, after the action of a hypotonic solution and fixation, consists of obtaining the germinal cell suspension by immersing pieces of tubules in a 60% acetic acid solution. Spreading and staining can be done as previously described.

2.7112 Culture Method

The first attempts at culture were made by Lima de Faria et al (1966), studying the replication of DNA in germinal cells. Dutrillaux (1971) devised a germinal cell culture technique that has been used in the field of male infertility, but its main purpose was to evaluate the in vitro effects of chemical or physical mutagens on sex cells.

With this technique, the seminiferous tubules are dilacerated and the cells are suspended. The suspension is transferred to a sterile tube and left in an incubator for approximately 48 hours at a temperature slightly lower than 37°C. Subsequent steps, such as hypotonic treatment, fixation, and preparation of slides, are similar to direct-examination cell suspension techniques.

2.712 Chromosomes from Ovarian Material

The study of oocytes at metaphases I and II developed only after the discovery of in vitro maturation techniques of oocytes during the preovulation phase of the menstrual cycle.

The main contributors to these techniques were Edwards (1965a and b; 1970), Tarkowski (1966), Jagiello et al (1968), Yuncken (1968), and Streptoe and Edward (1970) who helped codify the use of gonadotropins to stimulate follicular maturation and culminate in superovulation.

Ovaries or fragments of ovaries, according to the Edwards method (1965a and b), were placed in a saline solution containing 1 U of heparin per milliliter. The follicles were dissected under a microscope, transferred to the culture medium, and punctured to free the oocytes. The oocytes were placed in a petri dish containing a 199 medium with 15% fetal calf serum and antibiotics, and cultured in a 5% CO_2 atmosphere. The oocytes were examined at various times after beginning the culture, depending on the desired stage: metaphase I (28 to 35 hours) or metaphase II (36 to 43 hours). At the end of the culture period, the oocytes were submitted to the action of a hypotonic agent, fixed, stained, and examined. Excellent meiotic preparations have thus been shown by Yuncken (1968) and Edwards (1970).

The recent development of IVF programs has provided unfertilized oocytes. Oocytes are collected after follicular stimulation. These oocytes are considered unfertilized if they do not cleave within 48 hours after insemination. Since 1985, several cytogenetic studies have been reported (Michelmann and Mettler, 1985; Spielmann et al, 1985; Martin et al, 1986; Pellestor and Sele, 1988). Basically, the technique used to obtain metaphase chromosomes is that described by Tarkowski in 1966. Each oocyte is placed in 1% sodium citrate hypotonic solution at 37°C for 5 minutes, and swelling is observed under an inverted microscope. The oocyte is then transferred to a grease-free slide and fixed using 4 to 6 drops of fresh fixative (1 volume of acetic acid = 3 volumes of ethanol 95%). Chromosomes are identified on the slide by means of a phase-contrast microscope, and R-banding can be performed 24 hours later (Pellestor and Sele, 1988) (Figure 2.712.1).

2.72   INVESTIGATION OF PACHYTENE STAGE

2.721   Chromomere Technique and Obtaining Pachytene Bivalent Karyotypes from Human Spermatocytes

In contrast to metaphase I and II, obtaining well-spread pachytene bivalents extended over many years, from 1949, the date of the first attempt by Schultz and St Lawrence, to 1975, the date the first complete karyotype was established with the use of G-banding technique (Luciani et al, 1975). By the same time, several observations showed the close correspondence between mitotic/meiotic G-bands and chromomeres, structures spontaneously visible along pachytene bivalents (Hungerford et al, 1971a and b; Ferguson-Smith and Page, 1973; Luciani et al, 1975). More recently, decisive progress has been made in obtaining both well-spread pachytene bivalents and preservation of their chromomere structure following the introduction of prolonged hypotonic treatment (Luciani et al, 1975, 1984; Jagiello and Fang, 1982; Jhanwar et al, 1982). This technique, known as the chromomere technique, permits suitable pachytene karyotypes to be obtained, each autosomal bivalent being identified according to the number and sequence of their chromomeres (Figure 2.721.1a).

Immediately after removal, the testicular fragments are immersed in 10 ml of 0.88% KCl and kept at room temperature for 8 to 10 hours. The samples are transferred in fixative (methanol:acetic acid; 3:1 volume) and fixed for 12 to 18 hours, usually overnight, at room temperature. The next day, the fragments are shredded into the fixative solution. The cell suspension is pipetted into a conical vial and centrifuged at 800 rpm for 7 minutes. The pellet is resuspended in 5 ml of 45% acetic acid and immediately centrifuged at 800 rpm for 5 min-

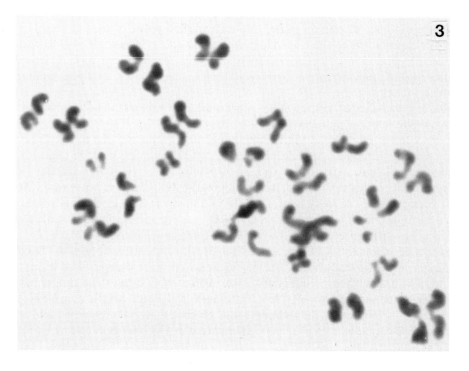

FIGURE 2.712.1. R-banded chromosome spread of a matrice oocyte at metaphase II stage (23,X) (magnification ×1500). (Courtesy of Pr. Sele, Grenoble.)

utes. The supernatant is discarded, and the pellet is suspended in 5 ml of fixative. The suspension is centrifuged for 5 minutes at 800 rpm. The major part of the supernatant is eliminated, the pellet is resuspended, and spreads can be made on clean precooled slides. The preparations are stained with 4% Giemsa solution and diluted in distilled water. The pH is adjusted to 6.7 with sodium phosphate buffer. This technique gives good results (Figure 2.721.1a), is very simple, and allows mailing of the samples (Guichaoua et al, 1986; Luciani et al, 1987). Various cytochemical staining techniques, such as centromeric heterochromatin visualization (Figure 2.721.1b) and in situ hybridization, can be performed from preparations with this technique.

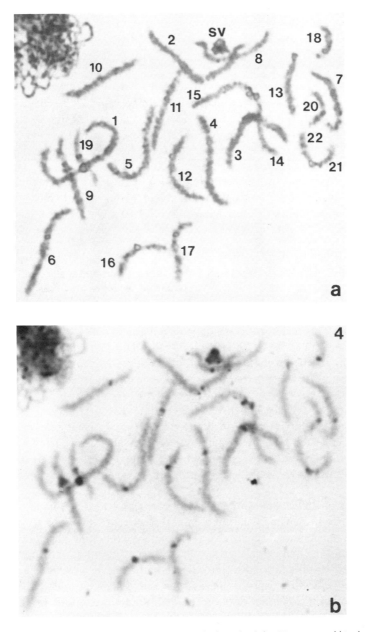

FIGURE 2.721.1a and b. Human pachytene spermatocyte in which each of the 22 autosomal bivalents can be identified on the basis of its chromomere pattern (a) and centromere position following C-banding (b). SV = sex vesicle (magnification ×1500).

2.722   Treatment of Fetal Ovaries and
Obtaining Pachytene Bivalents

All beginning stages of meiosis, including diplotene, can be studied using fetal ovaries. Squashing methods were first used most commonly (Ohno et al, 1962; Baker, 1963; Manotaya and Potter, 1963), but the best results have been obtained by cell suspension with air drying, initiated in our laboratory (Luciani et al, 1974).

Immediately after removal, the ovary is placed in fixative (methanol:glacial acetic acid = 3:1 volume) and can be stored for 24 or 48 hours. After fixation the ovary is cut into tiny pieces. The cell suspension and fragments obtained are treated separately.

1. All fragments are transferred to a conical test tube. Excess fixative is removed, and 10 drops of freshly prepared 45% acetic acid are added and thoroughly mixed. After 3 or 4 minutes, the germinal cells fall into suspension. The suspension is centrifuged at 800 rpm for 5 minutes. The supernatant is discarded, and the pellet is resuspended in a few drops of fixative. The germinal cells in suspension can now be dropped onto cold, wet slides and left to dry.

2. The cell suspension is centrifuged at 800 rpm for 8 minutes. The supernatant is removed, and the pellet is resuspended in a few drops of freshly prepared 45% glacial acetic acid. The suspension is thoroughly mixed and pipetted. Droplets are transferred to either precooled or dry slides. In the latter case, it is necessary to add some droplets of methanol to the final suspension to ensure good spreading. Finally, the preparations are stained with Giemsa solution or some other chromatin stain.

Apart from the ease and speed of handling, main interest in the procedure lies in the mailing of samples. This allows interesting studies about chromosome behavior at the pachytene stage in various abnormalities (Luciani et al, 1976, 1978a and b).

2.723   Microspreading Method for Study of Synaptonemal
Complexes by Light and Electron Microscopy (Figure 2.723.1)

The first whole-mount electron microscopy (EM) technique to study synaptonemal complexes (SCs) in the spermatocytes of the grasshopper, *Locusta migratoria*, was developed by Counce and Meyer (1973). This technique was adapted to mammals including human by Moses and his collaborators (1974, 1975, 1976, 1977a and b, 1977). Latter, Fletcher (1979) and Pathak and Elder (1980) published techniques to study SCs using light microscopy. Sequential study of SCs by light and electron microscopy is possible using the method described by Dresser and Moses (1980), Solari (1980), and Navarro et al (1981). Presently, we described a slightly modified technique of the method of Navarro et al (1981).

Sampling: Samples are obtained during testicular biopsy using general anesthesia. They are immediately collected in Ham F10 and 20% fetal calf serum. The technique is preferably realized during the hours after sampling. It is now possible to operate up to 2 days after biopsy, permitting the samples to be transported when they are obtained far from the laboratory.

Preparation of the cell suspension: The tubules are torn with fine forceps in Ham F10 without calf serum. The suspension is left in a conical glass centrifuge tube for about 20 minutes to allow larger fragments to sediment. The supernatant containing pachytene cells is changed to another conical tube and centrifuged at 800 rpm for 7 minutes. The pellet is washed 3 to 4 times with Ham F10 medium until the supernatant is clear.

Preparation of slides: For optical microscopic analysis, slides must be very clean and are prepared as for other meiotic laboratory techniques. Slides are dipped in chlorhydric acid for 30 minutes, rinsed with double-distilled water, dipped in 100 ethanol for 30 minutes, and dried with a paper napkin.

For electron microscopic analysis, the slides are coated with a viscous alkaline soap (house-

FIGURE 2.723.1. Electron micrograph of a surface-spread human spermatocyte at pachytene stage. The 22 complete autosomal synaptonemal complexes can be traced along their entire lengths. X and Y chromosomes are condensed in a convoluted structure. (Courtesy of Pr. Rumpler, Strasbourg.)

hold soap dissolved in water), air dried for 24 hours, and cleaned with a paper napkin until clear. The slides are now dipped for up to three quarters of their length either in an 0.5% formvar solution in chloroform kept at room temperature or in 2% collodion in isoamyl acetate preserved at 4°C. The slides are air dried in the vertical position. To prevent the coat of formvar or collodion from being removed during subsequent stages of the technique, the slide borders are sealed with nail polish. The slides are then dipped for about 10 seconds in an 0.01% solution of cytochrome C in sterile distilled water at 4°C and slowly removed so that the film is uniform. Coated slides are air dried. Prepared slides with formvar can be kept up to 1 month at room temperature; slides with collodion are kept at 4°C.

Spreading the cells and preparing specimen slides: An 0.2 M sucrose (3.4 g/50 ml) solution in double-distilled water is prepared and filtered through a Millipore filter. Several drops of the sucrose solution are placed in the same spot on a parafilm surface to obtain a bubble; a drop of the cell suspension is carefully expressed and gently touched to the surface of the spreading solution. The spread is allowed to stabilize for 1 or 2 minutes, and a slide is strongly touched to the surface of the bubble. A new bubble of spreading solution with a drop of cell suspension touched on is used for each slide. After the spread cells are picked up, the slides are immediately turned, laid flat on a support, and fixed.

Fixation: Either formaldehyde or paraformaldehyde can be used.

• A formaldehyde 9% solution in bidistilled water is prepared from the formaldehyde 37% solution of Merck.

- Preparation of paraformaldehyde: prepare paraformaldehyde 4% in 0.2 M sucrose solution. Heat slowly to 60° to 80°C while stirring and add about 6 drops of 1 N sodium hydroxide (NaOH) for 100 ml of solution until clear. Allow to cool, adjust to pH 9 to 10 with pH 9 borate buffer, and filter. The solution may be kept for a few days at 4°C. The slides are covered with the fixative for 8 to 10 minutes, washed for 30 seconds with an 0.4% solution of photoflow filtered just before use, and air dried.

Staining: It is better to wait 2 to 3 days before staining the slides, keeping the slides with collodion at 4°C. A 50% aqueous silver nitrate (AgNO₃) solution is prepared in double distilled water, and six drops per milliliter of a filtered 0.4% paraformaldehyde solution are added. Each slide is covered with 3 to 4 drops of staining solution and mounted with a coverslip. The slides are placed horizontally in a wet chamber and kept at 60°C. The time of staining varies from 30 to 60 minutes, and the color of the preparation is verified under the light microscope. When the slides are cold, they are washed with distilled water and air dried. The preparations are observed under the light microscope, and the cells selected and photographed. Their coordinates are recorded.

For electron microscopy, Moses (1977a) and Hulten et al (1986) stain the SCs with a 4% aqueous solution of phosphotungstic acid (PTA) filtered through a Millipore filter; this solution can be kept for a few days under refrigeration. The staining solution is made immediately before use by diluting one part of the 4% aqueous PTA with three parts of 95% ethanol. Slides are stained with ethanol-PTA for 15 minutes, washed in the same solvent, and air dried.

Transfer of selected pachytenes from slide to EM grids: The EM grids (75 mesh copper grids) are coated with a solution made by dissolving 2 cm of Scotch tape in 10 ml of chloroform to make them sticky. When the desired meiotic figure is localized under a light microscope, the lightfield is reduced so that its diameter becomes slightly greater than the diameter of a grid. The circumference of the lightfield is marked with four little spots. The same procedure is repeated for all the selected pachytene nuclei. With a sharp metal point, the plastic film is delicately torn by tracing a great rectangle that includes all the selected figures. The slide, with its specimen-carrying surface facing upwards, is slowly immersed in a distilled water trough at an angle of about 30° to the liquid surface. When the water reaches the lower side of the rectangle, the film strips off the slide and floats away. This is facilitated by the alkaline soap coating the slides. While the plastic film is floating on the water, the rough surface of the copper grids is deposited on it into the regions delimited by the spots. To pick up the grids with the floating collodion, filter paper is applied to the film, quickly pulled out when it is wet, and air dried. The grids are rapidly analyzed with the electron microscope.

2.724   Three-Dimensional Reconstruction of Zygotene and Pachytene Nuclei from Electron Microscope Serial Sections

Serial sectioning and reconstruction of whole nuclei have successfully been used to analyze zygotene and pachytene stages in a variety of organisms (Beyers and Goetsch, 1975; Gillies, 1972, 1973; Goldstein and Moens, 1976; Holm, 1977; Moens, 1973; Rasmussen, 1976). In man, the human pachytene karyotype, chromosome pairing, and recombination nodules have been studied by three-dimensional reconstruction (Holm and Rasmussen, 1977; Rasmussen and Holm, 1978; Bojko, 1983).

Electron microscope technique: Testicular biopsy specimens or fetal ovarian samples cut into 1 mm³ pieces are fixed at 4°C with 3% glutaraldehyde in 0.1 M phosphate buffer, pH 7.2, containing 1.5% sucrose for 20 minutes and postfixed at 4°C with 2% osmium tetroxide in the same buffer (without sucrose) for 15 minutes. The tissue is dehydrated in a graded acetone series. After embedding in epon, serial thin sections (100 nm) are picked up on single slot grids coated with formvar film. Sections are stained with uranylacetate and lead citrate.

Three-dimensional reconstruction: Analysis of one nucleus requires about 100 serial sections.

(a) For studying the bivalents at pachytene, the first step consists of identifying nuclei with synaptonemal complexes and a fully formed sex vesicle.

(b) For counting and localizing recombination nodules, in addition to pachytene, zygotene nuclei must be analyzed, because variations have been described (Rasmussen and Holm, 1978). At zygotene, the formation of the sex vesicle is not fully achieved; the lateral elements of the synaptonemal complexes are thinner than those at pachytene. The nucleolus is also useful in distinguishing the two stages: it has a reticulated (nucleolonemal) aspect at zygotene and becomes compact with segregation of its components at pachytene.

Several nuclei at their major diameter are selected from a section located in the middle of the series. The study progresses from this section towards the two ends of the series for each nucleus chosen.

Each nucleus is photographed using three magnifications: 2000×, 4000×, and 8000× or more. The smallest is useful to localize the nucleus within the seminiferous tubule or the ovary. A magnification of 4000 allows visualization of the entire nucleus and facilitates the assembly of structures such as synaptonemal complexes, nucleoli, and sex visicle. At a magnification of 8000, several micrographs are necessary to study one nucleus; details of bivalents and recombination nodules can be analyzed. Reconstruction is performed with the 8000× magnification (Figure 2.724.1).

Four guiding marks are chosen outside the nucleus (in the cytoplasm) on a micrograph from the middle of the series. A transparent plastic sheet is placed on the micrograph. The outlines of the nuclear envelope, mitochondria, vacuoles, and the guiding marks are drawn on this sheet.

This sheet is placed over the next micrograph to trace the guiding marks on it. Localiza-

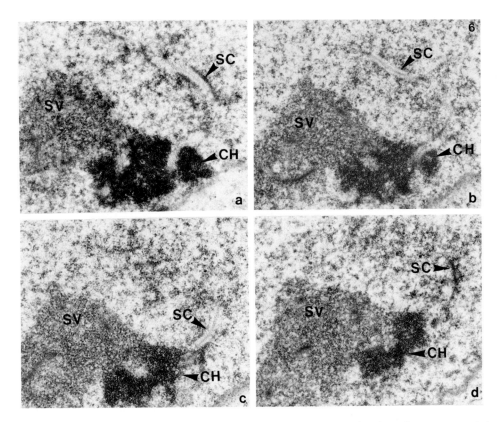

FIGURE 2.724.1a, b, c, and d. Four consecutive sections through an acrocentric bivalent whose centromeric heterochromatin (CH) is associated with the sex vesicle (SV). SC = synaptonemal complex (magnification ×14,280).

tion of these marks is facilitated by drawing the nuclear envelope and cytoplasmic structures on the sheet. On the following sheet the nuclear envelope and the marks are drawn from the preceding sheet, and so on up to the two ends of the series.

A new transparent sheet is taken on which the marks are localized. The synaptonemal complex corresponding to one bivalent is drawn from telomere to telomere, proceeding from one micrograph to another. In the same way, the 22 bivalents and the sex vesicle are traced one after another. When these tracings are completed, all bivalents are redrawn on a new transparent sheet, respecting their spatial relationships (Figure 2.724.2).

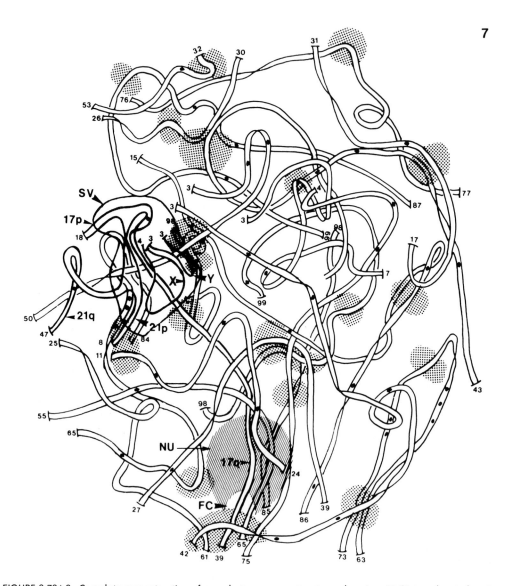

FIGURE 2.724.2. Complete reconstruction of a pachytene spermatocyte nucleus in a 17-21 translocated patient. Forty-two autosomes are paired into bivalents, each with a continuous synaptonemal complex from telomere to telomere. Normal and translocated chromosomes form a quadrivalent (17p and 21p: short arms of chromosomes 17 and 21; 17q and 21q: long arms of chromosomes 17 and 21 in the translocation figure). An asynapsis is observed in the region of breakpoints. Note the association between the quadrivalent and the sex vesicle: the unpaired region of the quadrivalent and paired regions are surrounded by the chromatin of the XY bivalent. Except for one free end on bivalent 1, all telomeres are attached to the nuclear envelope. Inside the synaptonemal complex dark points are the recombinational nodules. SV = sex vesicle; NU = nucleolus; FC = fibrillar center; dotted areas = centromeric heterochromatin.

ACKNOWLEDGMENTS: We wish to thank L. Laurens for typing the manuscript and M. Soler for photographic assistance. We are also very grateful to A. de Lanversin, F. de Fombelle, and M. F. Bertrand for their technical assistance.

2.8   REFERENCES

Arrighi FE, Hsu TC: Localization of heterochromatin in human chromosomes. *Cytogenetics* 1971;10:81–86.

Baker TG: A quantitative and cytological study of germ cells in human ovaries. *Proc Roy Soc B* 1963; 158:417–433.

Barnes DW, Sirbasku DA, Sato GH, eds: *Cell Culture Methods for Molecular and Cell Biology* (4 volumes). New York, Alan R. Liss, Inc., 1984.

Beyers B, Goetsch L: Electron microscopic observations on the meiotic karyotype of diploid and tetraploid Saccharomyces cerevisiae. *Proc Natl Acad Sci* 1975;72:5056–5060.

Bhatia R, Koppitch FC, Sokol RJ, Evans MI: Improving the yield direct chorionic villus slide preparation. *Am J Obstet Gynecol* 1986;154:408–411.

Blakemore KJ, Watson, MS, Samuelson J, Breg WR, Mahoney MJ: A method of processing first-trimester chorionic villous biopsies for cytogenetic analysis. *Am J Hum Genet* 1984;36:1386–1389.

Bojko M: Human meiosis VIII. Chromosome pairing and formation of the synaptonemal complex in oocytes. *Carlsberg Res Commun* 1983;48:457–483.

Brambati B, Oldrini A, Ferrazzi, Lanzani A: Chorionic villus sampling: An analysis of the obstetric experience of 100 cases. *Prenat Diagn* 1987;7:157–169.

Brambati B, Oldrini A: Methods of chorionic villus sampling, in Brambati B, Simoni G, Fabro S (eds): *Chorionic Villus Sampling*. New York, M. Dekker, 1986, pp 73-98.

Caspersson T, Hulten M, Lindsten J, Zech L: Identification of chromosome bivalents in human male meiosis by quinacrine mustard fluorescence analysis. *Hereditas (Lund)* 1971;67:147–149.

Cheung SW, Kyine M, Crane JP: First trimester chorionic villus biopsy – a method for obtaining 500 band resolution from direct chromosome preparations. *Am J Hum Genet* 1985;37:A88.

Counce SJ, Meyer GF: Differentiation of the synaptonemal complex and the kinetochore in *Locusta* spermatocytes studied by whole mount electron microscopy. *Chromosoma* 1973;44:231–253.

Dresser ME, Moses MJ: Synaptonemal complex karyotyping in spermatocytes of the Chinese hamster (*Cricetulus griseus*). IV. Light and electron microscopy of synapsis and nucleolar development by silver staining. *Chromosoma* 1980;76:1–22.

Dutrillaux B: La culture de cellules germinales mâles: méthodes et applications. *Ann Génét* 1971;14: 157–159.

Edwards RG: Maturation in vitro of human ovarian oocytes. *Lancet* 1965a;2:926–929.

Edwards RG: Maturation in vitro of mouse, sheep, cow, pig, rhesus monkey and human ovarian oocytes. *Nature (Lond.)* 1965b;208:354–349.

Edwards RG: Observations on meiosis in normal males and females, in Jacobs PA, Price WH, Law P (eds): *Human Population Cytogenetics*, Edinburgh, University Press, Pfizer Medical Monographs 5, 1970, pp 9–21.

Evans EP, Breckon G, Ford CE: An air-drying method for meiotic preparations from mammalian testes. *Cytogenetics* 1964;3:289–294.

Evans EP, Burtenshaw MD, Ford CE: Chromosomes of mouse embryos and newborn young: Preparations from membranes and tail tips. *Stain Technol* 1972;47:229–234.

Ferguson-Smith MA: The sites of nucleolus formation in human pachytene chromosomes. *Cytogenetics* 1964;3:124–134.

Ferguson-Smith MA, Page BM: Pachytene analysis in human reciprocal (10;11) translocation. *J Med Genet* 1973;10:282–287.

Filkins K, Russo JF (ed): *Human Prenatal Diagnosis*. New York, Marcel Dekker, Inc., 1985.

Fletcher JM: Light microscope analysis of meiotic prophase chromosomes by silver staining. *Chromosoma* 1979;72:241–248.

Flori E, Nisad I, Flori J, Dellenbach P, Ruch JV: Direct fetal chromosome studies from chorionic villi. *Prenat Diagn* 1985;5:287–289.

Ford CE, Evans EP: Meiotic preparations from mammalian testes, *in* Benirschke K (ed): *Comparative Mammalian Cytogenetics*, Intern. Conf. at Dartmouth Med. School. Hanover, New Hampshire, 1968. Berlin, Heidelberg, New York, Springer Verlag, 1969, pp 461–464.

Ford CE, Hamerton JL: The chromosomes of man. *Nature (Lond)* 1956;178:1020–1023.

Freshney RI: Culture of animal cells. *A Manual of Basic Techniques*. New York, Alan R. Liss, Inc., 1983.

Gagne R, Vagner-Capodano AM, Devictor-Vuillet M: Intérêt de la localisation de l'hétérochromatine pour l'identification des chromosomes méiotiques à la diacinèse chez l'homme. *CR Acad Sci (Paris)* 1973;276:769–771.

Gibas LM, Grujic S, Barr MA, Jackson LG: A simple technique for obtaining high quality chromosome preparations from chorionic villus samples using FdU synchronization. *Prenat Diagn* 1987; 7:323–327.

Gibas M, Gibas Z, Sandberg AA: Technical aspects of cytogenetic analysis of human solid tumors. *Karyogram* 1984;10:25–27.

Gillies CB: Reconstruction of the Neurospora crassa pachytene karyotype from serial sections of synaptonemal complexes. *Chromosoma (Berl)* 1972;36:119–130.

Gillies CB: Ultrastructural analysis of maize pachytene karyotypes by three dimensional reconstructions of the synaptonemal complexes. *Chromosoma (Berl)* 1973;43:145–176.

Goldstein P, Moens PB: Karyotype analysis of Ascaris lumbricoides var. *suum Chromosoma (Berl)* 1976;58:101–111.

Gregson NM, Seabright M: Handling chorionic villi for direct chromosome studies. *Lancet* 1983;II:1491.

Guichaoua MR, Delafontaine D, Taurelle R, Taillemite JL, Morazzani MR, Luciani JM: Loop formation and synaptic adjustment in a human male heterozygous for two pericentric inversions. *Chromosoma (Berl)* 1986;93:313–320.

Hahnemann N: Early prenatal diagnosis: A study of biopsy techniques and cell culturing from extraembryonic membranes. *Clin Genet* 1974;6:294–306.

Hogge WA, Schonberg SA, Golbus MS: Chorion villus sampling: Experience of the first 1000 cases. *Am J Obstet Gynecol* 1986;154:1249–1252.

Hogge WA, Schonberg SA, Golbus MS: Prenatal diagnosis by chorionic villus sampling: Lessons of the first 600 cases. *Prenat Diagn* 1985;5:393–400.

Holm PB: Three-dimensional reconstruction of chromosome pairing during the zygotene stage of meiosis in Lilium longiflorum (Thunb.). *Carlsberg Res Commun* 1977;42:103–151.

Holm PB, Rasmussen SW: Human meiosis I. The human pachytene karyotype analyzed by three dimensional reconstruction of the synaptonemal complex. *Carlsberg Res Commun* 1977;42:283–323.

Hulten M: Chiasma distribution at diakinesis in the normal human male. *Hereditas* 1974;76:55–78.

Hulten M, Lindsten J, Ming-Pen-Ming L, Fraccaro M: The XY bivalent in human male meiosis. *Ann Hum Genet* 1966;30:119–123.

Hulten MA, Saadallah N, Wallace BMN, Creasy MR: Meiotic studies in man. Human cytogenetics, a practical approach, *in* Rooney BE, Czepulkowski HH (eds): Published in the practical approach series. Rickwood D, Hames BD (series eds). IRC Press Oxford, Washington DC, 1986, pp 163–196.

Hungerford DA, La Badie GU, Balaban GB: Chromosome structure and function in man. II. Provisional maps of the two smallest autosomes (chromosomes 21 and 22) at pachytene in the male. *Cytogenetics* 1971a;10:33–37.

Hungerford DA, La Badie GU, Balaban GB, Messatzzia LR, Haller G, Miller AE: Chromosome structure and function in man. IV. Provisional maps of the three long acrocentric autosomes (chromosomes 13, 14 and 15) at pachytene in the male. *Ann Génét* 1971b;14:257–260.

Jackson L: CVS mosaicism. Abstract of the 3rd International Conference on CVS, Chicago, 1987.

Jackson LG: Prenatal genetic diagnosis by chorionic villus sampling, in Porter IH, Hatcher NH, Willey AM (eds): *Perinatal Genetics: Diagnosis and Treatment*, New York, Academic Press, 1986, pp 95–113.

Jackson LG, Wapner RA, Barr MA: Safety of chorionic villus biopsy. *Lancet* 1986;I:674–675.

Jackson LG, Wapner RJ: Risks of chorion villus sampling. *Ballier's Clin Obstet Gynecol* 1987;1:513–531.

Jagiello GM, Karnicki J, Ryan RJ: Superovulation with pituitary gonadotrophins. Method for obtaining meiotic metaphase figures in human ova. *Lancet* 1968;1:178–180.

Jagiello GM, Fang JS: Complete autosomal chromomere maps of human early and mid/late pachytene spermatocytes. *Am J Hum Genet* 1982;34:112–124.

Jhanwar SC, Burns JP, Alonso ML, Hew W, Chaganti RSK: Mid-pachytene chromomere maps of human autosomes. *Cytogenet Cell Genet* 1982;33:240–248.

Karson EM, Dorfmann AD, Landsberger EJ, Schulman JD, Larsen JW, Evans MI: Modifications of media, incubation, and fixation for direct preparation of chorionic villi. *Am J Hum Genet* 1984; 36:191S.

Kazy Z, Bakharev VA, Stygar AM: Value of the ultrasonic studies in biopsy of the chorion, according to genetic indicators. *Akush-Ginekol (Mosk)* 1979;8:29–31.

Kazy Z, Stygar AM, Bakharev VA: Chorionic biopsy under immediate realtime (ultrasonic) control. *Orv-Hetil* 1980;121:2765–2766.

Kazy Z, Demidov VN, Stygar AM, Bakharev VA: Use of realtime scanning for chorionic biopsy. *Akush:Ginekol* 1981;3:51–54.

Kazy Z, Rozovsky IS, Bakharev VA: Chorion biopsy in early pregnancy: A method of early prenatal diagnosis for inherited disorders. *Prenat Diagn* 1982;2:39–45.

Kruse PF Jr, Patterson MK: *Tissue Culture. Methods and Application.* New York, Academic Press, 1973.

Ledbetter S, Ledbetter DL: Personal communication, 1985–1987.

Leibovitz A, Liu R, Hayes C, Salmon SE: A hypoosmotic medium disaggregate tumor cell clumps into viable and clonogenic single cells for the human tumor stem cell clonogenic assay. *Int J Cell Cloning* 1983;1:478–485.

Leibovitz A: Development of tumor cell lines. *Cancer Genet Cytogenet* 1986;19:11–19.

Lima De Faria A, German J, Ghatnekar M, McGovern J, Anderson L: DNA synthesis in the meiotic chromosomes of man. A preliminary report. *Hereditas (Lund)* 1966;56:398–399.

Limon J, Cin PD, Sandberg AA: Application of long-term collagenase disaggregation for the cytogenetic analysis of human solid tumors. *Cancer Genet Cytogenet* 1986;23:305–313.

Luciani JM, Devictor-Vuillet M, Stahl A: Hypotonic KCl: An improved method of processing human testicular tissue for meiotic chromosomes. *Clin Genet* 1971;2:32–36.

Luciani JM, Devictor-Vuillet M, Gagne R, Stahl A: An air-drying method for first meiotic prophase preparations from mammalian ovaries. *J Reprod Fertil* 1974;36:409–411.

Luciani JM, Morazzani MR, Stahl A: Identification of pachytene bivalents in human male meiosis using G-banding technique. *Chromosoma* 1975;52:275–282.

Luciani JM, Devictor M, Morazzani MR, Stahl A: Meiosis of trisomy 21 in the human pachytene oocyte. *Chromosoma* 1976;57:155–163.

Luciani JM, Devictor M, Boue J, Freund M, Stahl A: The meiotic behaviour of triploidy in a human 69,XXX fetus. *Cytogenet Cell Genet* 1978a;20:226–231.

Luciani JM, Devictor M, Boue J, Morazzani MR, Stahl A: Etude de la méiose ovocytaire chez un foetus trisomique 18. Comportement du chromosome surnuméraire et identification du bivalent 18. *Ann Génét (Paris)* 1978b;21:215–218.

Luciani JM, Guichaoua MR, Morazzani MR: Complete pachytene chrommomere karyotypes of human spermatocyte bivalents. *Hum Genet* 1984;66:267–271.

Luciani JM, Guichaoua MR, Delafontaine D, North MO, Gabriel-Robez O, Rumpler Y: Pachytene analysis in a 17;21 reciprocal translocation carrier: Role of the acrocentric chromosomes in male sterility. *Hum Genet* 1987;77:246–250.

Macera MJ, Szabo P, Verma RS: An improved method for short term culturing of bone marrow from acute lymphocytic leukemia (ALL) and acute myelogenous leukemia (AML). Submitted, 1988.

Makino S, Nishimura I: Water-pretreatment squash technic. *Stain Technol* 1952;27:1–7.

Mandahl N, Gustavii B, Heim S, Kristoffersson U, Mineur A, Mitelman F: Technical aspects of long-term culture and cytogenetic analysis of first trimester chorionic villi. *Karyogram* 1985;11:10–13.

Manotaya T, Potter EL: Oocytes in prophase of meiosis from squash preparations of human fetal ovaries. *Fertil & Steril* 1963;14:378–392.

Martin RH, Mahadevan MM, Taylor PJ, Hildebrand K, Long-Simpson L, Peterson D, Yamamoto J, Fleetham J: Chromosomal analysis of unfertilized human oocytes. *J Reprod Fert* 1986;78:673–678.

McBain JA, Weese JL, Meisner LF, Wolberg WH, Wilson JKV: Establishment and characterization of human colorectal cancer cell lines. *Cancer Res* 1984;44:5813–5821.

Meredith R: A simple method for preparing meiotic chromosomes from mammalian testis. *Chromosoma (Basel)* 1969;26:254–258.

Metaxotou C, Antsaklis A, Panagiotopoulou P, Benetou M, Mavrou A, Matsaniotis N: Prenatal diagnosis of chromosomal abnormalities from chorionic biopsy samples: Improved success rate using a modified direct method. *Prenat Daign* 1987;7:461–469.

Michelmann HW, Mettler L: Cytogenetic investigations on human oocytes and early human embryonic stages. *Fertil & Steril* 1985;43:320–322.

Milunksy A (ed): *Genetic Disorder and Fetus.* New York, Plenum Press, 1988.

Moens PB: Quantitative electron microscopy of chromosome organization at meiotic prophase. Cold Springs Harbor Symp Quant Biol 1973;38:99–107.

Mohr J: Fetal genetic diagnosis: Development to techniques for early sampling of foetal cells. *Acta Pathol Microbiol Scand* 1968;73:73–77.

Moorhead PS, Nowell PC, Mellman WJ, et al: Chromosome preparations of leukocytes cultured from human peripheral blood. *Exp Cell Res* 1960;20:613–616.

Moses JM, Counce SJ: Synaptonemal complex karyotyping in spreads of mammalian spermatocytes, *in* Grell R (ed): Symposium on Mechanisms in Recombination. New York, Plenum Press, 1974, pp 385–390.

Moses MJ, Counce SJ, Paulson DF: Synaptonemal complex complement of man in spreads of spermatocytes with details of the sex chromosome pair. *Science* 1975;187:363–365.

Moses MJ, Counce SJ: Analysis of synaptonemal complex karyotypes in five mammals. *J Cell Biol* 1976; 70:131a.

Moses MJ: Synaptonemal complex karyotyping in spermatocytes of the Chinese hamster (*Cricetulus griseus*). I. Morphology of the autosomal complement in spread preparations. *Chromosoma* 1977a;60:99–125.

Moses MJ: Synaptonemal complex karyotyping in spermatocytes of the Chinese hamster (*Cricetulus griseus*). II. Morphology of the XY pair in spread preparations. *Chromosoma* 1977b;60:127–137.

Moses MJ, Slatton GH, Gambling TM, Starmer CF: Synaptonemal complex karyotyping in spermato-cytes of the Chinese hamster (*Cricetulus griseus*). III. Quantitative evaluation. *Chromosoma* 1977; 60:345–375.

Navarro J, Vidal F, Guitart M, Egozcue J: A method for the sequential study of synaptonemal com-plexes by light and electron microscopy. *Hum Genet* 1981;59:419–421.

Niazi M, Coleman DV, Loeffler FE: Trophoblast sampling in early pregnancy. Culture of rapidly divid-ing cells from immature placental villi. *Br J Obstet Gynaecol* 1981;88:1081–1085.

Ohno S, Klinger HP, Atkin NB: Human oogenesis. *Cytogenetics* 1962;1:42–51.

Pathak S, Elder FFB: Silver stained accessory structures on human sex chromosomes. *Hum Genet* 1980; 54:171–175.

Pearson PL, Bobrow M: Definitive evidence for the short arm of the Y chromosome associating with the X chromosome during meiosis in the human male. *Nature (Lond)* 1970;226:959.

Pellestor F, Sele B: Assessment of aneuploidy in the human female by using cytogenetics of IVF failures. *Am J Hum Genet* 1988;42:274–283.

Pergament E, Verlinsky Y, Chu L, Rafi KS, Shafer DA: Synchronization studies with chorionic villi. *Am J Hum Genet* 1984;37:A224.

Pitmon D, Exterman P, Engel E: Simplified chromosome preparations from chorionic villi obtained by choriocentesis or derived from induced abortions. *Ann Genet* 1984;27:254–256.

Rasmussen SW: The meiotic prophase in Bombyx mori females analyzed by three dimensional recon-structions of synaptonemal complexes. *Chromosoma (Berl)* 1976;54:245–293.

Rasmussen SW, Holm PB: Human meiosis II. Chromosome pairing and recombination nodules in human spermatocytes. *Carlsberg Res Commun* 1978;43:275–327.

Schultz J, St Lawrence P: A cytological basis for a map of the nucleolar chromosome in man. *J Hered* 1949;40:31–38.

Schwab ME, Muller C, Schmid-Tannwald I: Fast and reliable culture method for cells from 8-10 week trophoblast tissue. *Lancet* 1984;1:1082.

Simoni G, Brambati B, Danesino C, Terzoli GL, Romitti L, Rosella F, Fraccaro M: Diagnositic appli-cation of first trimester trophoblast sampling in 100 pregnancies. *Hum Genet* 1984;66:252–259.

Simoni G, Brambati B, Danesino C, Rossella F, Terzoli GL, Gerrari M, Fraccaro M: Efficient direct chromosome analyses and enzyme determinations from chorionic villi samples in the first trimester of pregnancy. *Hum Genet* 1983;63:349–357.

Simoni G, Gimelli G, Cuoco C, Romitti L, Terzoli G, Guerneri S, Rossella F, Pescetto L, Pezzplo A, Porta S, Brambati B, Porro E, Fraccoro M: First trimester fetal karyotyping: One thousand diag-noses. *Hum Genet* 1986;72:203–209.

Solari AJ: Synaptonemal complexes and associated structures in microspread human spermatocytes. *Chromosoma* 1980;81:315–337.

Spielmann HC, Kruger C, Stauber M, Vogel R: Abnormal chromosome behavior in human oocytes which remained unfertilized during in vitro fertilization. *J Ivf Et* 1985;2:138–142.

Streptoe PC, Edwards RG: Laparoscopic recovery of preovulatory human oocytes after priming of ova-ries gonadotrophins. *Lancet* 1970;1:683–690.

Tarkowski AK: An air-drying method for chromosome preparations from mouse eggs. *Cytogenetics* 1966;5:394–400.

Terzoli GL, Lalatta F, Gilbert F: Chorionic villus sampling: Improved method for operation of karyo-types after short-term incubation. *Prenat Diagn* 1987;7:389–394.

Tjio JH, Levan A: The chromosome number of man. *Hereditas (Lund)* 1956;42:1–6.

USNICHD Study Group. Midtrimester amniocentesis for prenatal diagnosis. Safety and accuracy. *JAMA* 1976;234:1471–1476.

Ward RHT, Modell B, Petrou M, Karagozlu F, Douratsos E: A method of chorionic villus sampling in the first trimester of pregnancy under real time ultrasonic guidance. *Br Med J* 1983;286:1542–1544.

Yuncken C: Meiosis in the human female. *Cytogenetics* 1968;7:234–238.

3

Banding Techniques

The past two decades have witnessed a revolution in our understanding of the human genome. Proliferation of newer staining techniques has facilitated precise identification of individual chromosomes, bringing clinical cytogenetics to the brink of an exciting new era. The difficulties in obtaining satisfactory bands have been frustrating to many technical personnel even today. Improved resolution of chromosome bands has resulted in the publication of over a dozen modifications of the available banding techniques. Nevertheless, one must search the literature for a method that suits an individual set of laboratory conditions, but this wastes time. The mystical factors associated with satisfactory and reproducible results remain unexplained. Thus, we will describe step by step some of the procedures that in our experience have produced consistent results. Furthermore, the steps in which variation may be required are indicated. We will attempt to elucidate the basic mechanisms involved in each banding in order to provide the reader with a better understanding.

3.1 DIFFERENTIAL STAINING TECHNIQUES

Differential staining techniques include the fluorescent and Giemsa staining methods that induce the light and dark bands along the length of chromosomes. These methods, since their inception, have been indispensable in the unequivocal identification of chromosomes.

3.11 Q-BANDS BY FLUORESCENCE USING QUINACRINE (QFQ)

The first major breakthrough in longitudinal differentiation of the human chromosomes came from Caspersson and his colleagues who discovered that chromosomes can be stained with quinacrine mustard to observe bright and dull fluorescence gradations using ultraviolet light. This method was termed as the Q-banding technique (Paris Conference, 1971). Later, a more descriptive term, QFQ technique (Q-bands by fluorescence using quinacrine), was applied to the method (ISCN, 1978).

Protocol 3.11.1   QFQ Technique

Solutions

1. McIlvaine's buffer (pH 5.4)

Citric acid ($H_3C_6H_5O_7$	2.1 g
Sodium phosphate dibasic (Na_2HPO_4)	3.9 g
Distilled water	500 ml

Instead of sodium phosphate dibasic (Na_2HPO_4), 7.45 g of sodium phosphate dibasic, 7 hydrate ($Na_2HPO_4 \cdot 7H_2O$) or 9.97 g of sodium phosphate dibasic, 12 hydrate ($Na_2HPO_4 \cdot 12H_2O$) can be used to prepare the same amount of buffer.

2. Quinacrine staining solution

    Atabrine   1 tablet or 100 mg

    McIlvaine's buffer (pH 5.4)   200 ml

Dissolve thoroughly and filter. Store in the dark at 2° to 5°C.

Procedure

1. Stain the slide in quinacrine staining solution for 10 to 15 minutes in the dark.
2. Rinse the slide in running tap water to remove excess quinacrine.
3. Place the slide in McIlvaine's buffer (pH 5.4) for 1 minute.
4. Mount the slide in the same buffer and blot off excess buffer.
5. Examine the slide under a fluorescence microscope using a wavelength of 450 to 500 nm.

Comments

Fluorescence intensity of chromosomes gradually fades away on exposure to ultraviolet light during microscopic observation. Exposure of metaphases to the light source should be kept to a minimum, especially if the slides are Q-banded to aid chromosome identification and are to be used subsequently for other staining procedures. The microscope objectives with annular diaphragm are particularly useful in fluorescence studies to regulate the intensity of light for obtaining optimum contrast and avoiding excessive light that would cause rapid fading. The cells should be photographed, if needed, before microscopic analysis. A high speed and high contrast panchromatic film, such as Kodak Tri X, is recommended for photographing fluorescence bands.

Staining Pattern

Chromosomes stained by quinacrine show a series of bright and dull fluorescent regions along the length. These bands are characteristic of each chromosome and help in their identification. The distal region in the long arm of the Y chromosome shows remarkably bright fluorescence and is the most intense in the entire complement (Figure 3.1.1). The pericentric regions of chromosomes 3 and 4, and the pericentric and satellite regions of acrocentric chromosomes may show significant variation, known as polymorphism or heteromorphism, in the fluorescence intensity between homologues as well as in different individuals.

Mechanism

Although the mechanism involved in differential fluorescence requires much more investigation, it is an accepted fact that bright fluorescence is found in chromosome regions known to consist of DNA rich in adenosine and thymidine (AT) base pairs (Ellison and Barr, 1972). Recent studies indicated that AT-rich DNA tends to enhance the fluorescence of quinacrine, whereas guanine and cytosine(GC)-rich DNA tends to quench this fluorescence (Comings et al, 1975; Comings, 1978; Miller et al, 1973). Furthermore, the variation in the base composition of DNA along the length of the chromosome may be the primary basis for Q-banding. Further evidence clearly indicates that protein-DNA interactions may also play a role in Q-banding (Pachmann and Rigler, 1972; Weisblum and deHaseth, 1973).

Applications

Q-banding not only is useful for routine chromosome identification, but also is one of the most elaborately used techniques to study the heteromorphisms associated with chromosomes 3 and 4, the acrocentric chromosomes, and the Y. Some of these heteromorphisms are used as chromosome markers to determine the parental origin of additional chromosomes in trisomies and in paternity studies. Q-banding is ideal for sequential banding in which chromosome identification is a prerequisite.

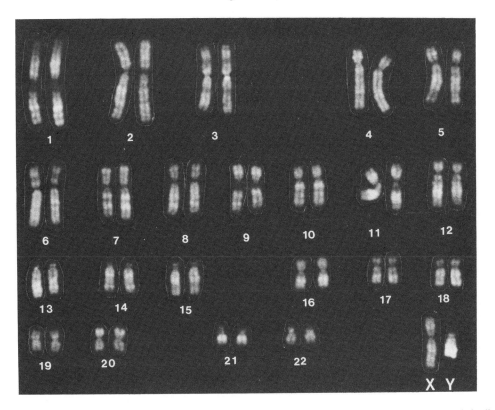

FIGURE 3.1.1. Karyotype of a male, demonstrating the fluorescence pattern by QFQ-technique (Q-bands by fluorescence using quinacrine).

3.12 G-BANDS BY TRYPSIN USING GIEMSA (GTG)

The Giemsa bands (G-bands) obtained by digesting the chromosomes with proteolytic enzyme trypsin are the most widely used in clinical laboratories for routine chromosome analysis. This technique is described as GTG (G-bands by trypsin using Giemsa) (ISCN, 1978).

Protocol 3.12.1 GTG Technique

Solutions

1. Dulbecco's phosphate-buffered saline-calcium and magnesium free (Dulbecco's PBS-CMF)
   As obtained from the media suppliers.

2. Trypsin solution (0.05%)

Trypsin (1:250)	50 mg
Dulbecco's PBS-CMF	100 ml

   Prepare fresh and use at room temperature for 3 to 4 hours. Discard the solution as soon as turbidity develops or the slides show an indication of contamination.

3. Giemsa staining solution (5%)

Distilled water	42.5 ml
Buffer solution (Gurr's, pH 6.8)	5.0 ml
Giemsa stain	2.5 ml

   The staining solution is made up fresh and used for 2 to 3 hours. The buffer solution (Gurr's, pH 6.8) is prepared by dissolving a Gurr's buffer tablet in 1 liter of distilled water.

Procedure

1. Age the air-dried preparations overnight at 55° to 60°C in an oven.

2. Treat the slides in trypsin solution (0.05%) for 5 to 10 seconds.

3. Rinse the slides briefly in cold Dulbecco's PBS-CMF (2° to 5°C, kept in the refrigerator).

4. Stain the treated slides in Giemsa staining solution for 4 to 6 minutes.

5. Rinse in distilled water and allow to dry.

6. The slides can be examined without mounting using dry objectives or they can be covered with a cover glass using permount for oil-immersion objectives.

Comments

The trypsin solution and Giemsa staining solutions should be changed periodically during the day to maintain optimum banding and staining conditions, respectively. A jar filled with Dulbecco's PBS-CMF is kept in the refrigerator and is taken out briefly to rinse the slide following trypsin digestion. All other solutions are left at room temperature.

Chromosome preparations obtained from different tissues respond with a different degree of sensitivity to trypsin digestion. Chromosomes prepared from fibroblasts, blood lymphocytes, and bone marrow react with an increasing order of sensitivity and therefore should be treated for shorter periods in trypsin solution, respectively.

Staining Pattern

Chromosomes stained by this protocol exhibit light and dark stained regions (light and dark bands) along the length of the chromosomes (Figure 3.1.2A & B). The G-band pattern induced by trypsin is grossly similar to that produced by quinacrine.

Applications

No special emphasis is needed and no description can justify the unlimited applications of G-bands. G-bands are routinely used for clinical evaluation of human chromosomes in most laboratories. The G-bands obtained by GTG are far superior in resolution to those using fluorochromes. The permanent nature of these preparations facilitates a thorough microscopic analysis of metaphases.

Mechanism

The mechanism of G-banding in not yet completely understood. A direct role of Giemsa stain in producing the G-bands has been suggested (McKay, 1973; Schuh et al, 1975). Comings (1978) suggests that G-banding occurs because metaphase chromosome contains a basic structure that is subjected to enhancement. How this enhancement occurs, however, is unclear. One theory states that Giemsa stain used in the GTG technique interacts primarily with chromosomal proteins (Daniel et al, 1973), although in vitro studies indicate that the only significant binding occurs with DNA (Comings and Avelino, 1975). Clark and his colleagues (Clark and Felsenfeld, 1975) suggest that arginine-rich histone is involved in GTG-banding, and there is no loss of either DNA or protein from the chromosome during trypsin treatment. It is hypothesized that differences between positive and negative G-bands may be due to the distribution of chromosomal protein and DNA. It is further suggested that positive G-bands are relatively rich in protein disulfides, whereas negative bands contain sulfhydryls instead. An intriguing aspect of G- and Q-banding is that although the banding patterns are similar, the mechanisms involved in producing these patterns are very different.

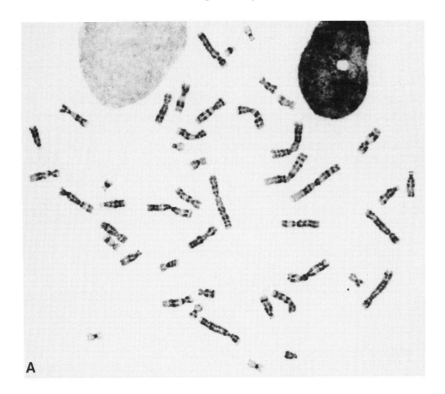

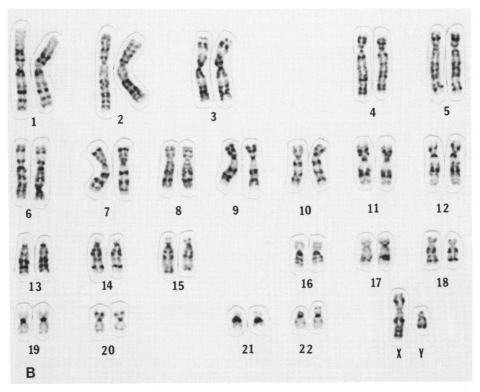

FIGURE 3.1.2. (A) Metaphase chromosomes of male fetus from an amniocyte cell culture, showing GTG-bands (G-bands by trypsin using Giemsa). (B) Karyotype of the same cell.

3.13 R-BANDS BY FLUORESCENCE USING ACRIDINE ORANGE (RFA) AND R-BANDS BY HEAT USING GIEMSA (RHG)

The basic principle behind the so-called reverse bands (R-bands) is the treatment of slides at high temperatures in various buffers followed by staining with either acridine orange (RFA) or Giemsa (R-bands by heat using Giemsa, RHG) (to be described in this section). This technique produces bands on human chromosomes that are the reverse of Q- and G-bands, that is, those areas that do not stain well by Q- or G-banding are intensely stained by the R-banding procedure (Dutrillaux and Lejeune, 1971; Sehested, 1974; Bobrow and Madan, 1973). Other methods of producing R-bands include staining of chromosomes with fluorochromes such chromomycin A_3 and olivomycin, and incorporation of a thymidine analogue, bromodeoxyuridine (BrdU), into replicating DNA during the terminal period of S-phase that could be elicited by a fluorochrome or by Giemsa. These methods are described later in respective sections on fluorescent stains and replication sequences.

Protocol 3.13.1   RFA Technique

Solutions

1. Sorensen's buffer (0.06 M, pH 6.5)

Potassium phosphate monobasic (KH_2PO_4)	5.60 g
Sodium phosphate dibasic (Na_2HPO_4)	2.64 g
Distilled water	1 L

Instead of sodium phosphate dibasic (Na_2HPO_4), 4.98 g of sodium phosphate dibasic, 7 hydrate ($Na_2HPO_4 \cdot 7H_2O$) or 6.66 g of sodium phosphate dibasic, 12 hydrate ($Na_2HPO_4 \cdot 12H_2O$) can be used to prepare the same amount of buffer.

2. Acridine orange staining solution (0.01%)

Acridine orange	10 mg
Sorensen's buffer (0.06 M, pH 6.5)	100 ml

Dissolve the acridine orange in buffer and filter. Store in the dark at 2° to 5°C.

Procedure

1. Age the slide at room temperature for 1 to 3 weeks. Fresh slides are not suitable.
2. Preheat Sorensen's buffer (0.06 M, pH 6.5) in a jar in a water bath at 85°C.
3. Incubate the slides in Sorensen's buffer at 85°C for 8 minutes for 1-week to 10-day-old slides.
4. Rinse the slides thoroughly in Sorensen's buffer at room temperature.
5. Stain in acridine orange staining solution (0.01%) for 4 to 5 minutes.
6. Rinse the slides in Sorensen's buffer for 1 minute.
7. Mount the slide in Sorensen's buffer and examine using a wavelength of 450 to 500 nm.

Comments

The duration of treatment in Sorensen's buffer at 85°C is critical and depends on the age of the slide. Treatment for 8 minutes is suitable to band the slides of 1-week to 10-day-old slides. Treatment should be shorter with older slides. The extent of acridine orange staining is also critical in obtaining optimum differentiation. If chromosomes appear too reddish, they are overstained, and the excess stain should be removed by rinsing repeatedly in Sorensen's buffer. Optimum staining shows various regions of chromosomes stained in different gradations between green and red. Photography can be performed using either a color film (Kodachrome 64) or a panchromatic high contrast film. Metaphases photographed using color film can be reproduced by printing on black and white photographic paper (Figure 3.1.3).

Protocol 3.13.2   RHG Technique

Solutions

1. Sorensen's buffer (0.06 M, pH 6.5)
   Same as in protocol 3.13.1.
2. Giemsa staining solution
   Same as in protocol 3.12.1.

Procedure

1. Treat the slides as described in protocol 3.13.1, steps 1 to 4.
2. Stain the slides in Giemsa staining solution until proper intensity is obtained (6 to 10 minutes).
3. Rinse in distilled water and allow to dry.
4. Mount in permount.

Comments

See comments under previous protocol. The slides are best observed using phase-contrast microscopy.

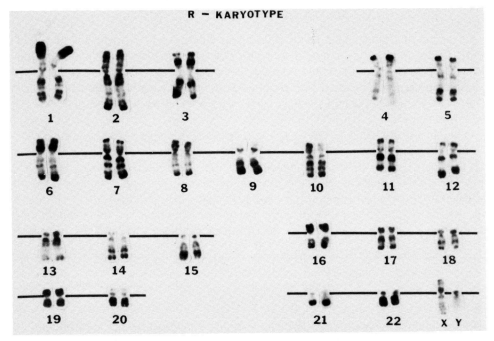

FIGURE 3.1.3. R-banded karyotype by RFA-technique (R-bands by fluorescence using acridine orange). Cell photographed with Kodachrome 64 and reproduced in black and white. (See text.)

Staining Pattern

Chromosomes stained by either of the foregoing protocols exhibit R-bands (Figure 3.1.3). The R-band regions by RBA are seen as green, whereas negative R-bands appear red, and the bands appear as dark and faint regions by RHG, a pattern that is the reverse of that observed by GTG (Dutrillaux and Lejeune, 1971; Verma and Lubs, 1975; 1976).

Applications

Several clinical laboratories prefer R-bands over G-bands for routine clinical cytogenetic investigation. One of the most important advantages of R-bands is that the telomeric regions of several chromosomes that are stained faintly in Q- and G-banding are stained dark.

Mechanism

The mechanism of reverse banding is not fully understood (Comings, 1978). Heat treatment induces denaturation of chromosomal proteins as well as of the AT-enriched nucleotide sequences in DNA, leaving the GC-rich DNA of the R-bands in a native configuration (Sumner, 1982). Selective denaturation of AT-rich DNA can be explored using acridine orange staining after heat treatment in various buffers (Burkholder, 1981). Consequently, acridine orange stains single-stranded DNA red and double-stranded DNA green (Rigler, 1966). Verification of the assumption that base composition is the major factor in R-banding comes from the use of fluorescent compounds such as chromomycin A_3, olivomycin, and mithramycin that preferentially bind to GC-rich DNA or show greater fluorescence in the presence of GC- than AT-rich DNA. It remains to be explored why R-bands can be seen with Giemsa staining. Comings (1978) believes that at high temperatures the AT-rich DNA of the G-bands is denatured and partially extracted, whereas the native DNA of the R-bands is not. Further comparison of the ultrastructure of G- and R-banded chromosomes indicates that the differences in electron density between band and interband regions is much more subtle in R-banded chromosomes than in G-banded ones (Burkholder, 1981).

3.2 SELECTIVE STAINING TECHNIQUES

The human chromosomes contain enormously large amounts of DNA which is stained highly specific. Recent progress in staining these regions by so-called "selective staining techniques" has been tremendous. Among them, the most recent is the application of restriction endonucleases. Chromosomes treated with endonucleases exhibit a specific characteristic pattern of DNA extraction with reduced staining, which leads researchers to believe that DNA extraction depends on the size of the DNA fragments produced by a particular enzyme, which is determined by the distance between two successive available enzyme-specific sites. It has been concluded that human genome is not just a linear display of bands along the arms, but chromatin does aggregate in an orderly manner with successive DNA folding producing selective bands. The rapid advances in staining methods resulted in the biologists' abilities to determine or inquire into the basic structure of these specific bands. The remarkable discovery that these bands are subject to considerable variation has further stimulated our understanding of the role of heterochromatin in the organization of human genome as well as their application in clinical cytogenetics. We have covered in some depth the most frequently used techniques in the laboratory.

3.21 C-BANDS

The C-banding technique produces selective staining of constitutive heterochromatin. These bands are mostly located at the centromeric regions of human chromosomes, hence they are known as C-bands. The original method described by Arrighi and Hsu (1971) involves primary treatment with an alkali, sodium hydroxide (NaOH), to denature the chromosomal

DNA and subsequent incubation in a salt solution. In a subsequent method described by Sumner (1972), a milder alkali, barium hydroxide ($Ba(OH)_2$), is used instead of NaOH. Both of these methods produce a similar characteristic pattern of C-bands. The latter method may provide an advantage, because it facilitates better control of a denaturation process.

Protocol 3.21.1 C-Bands Using Sodium Hydroxide (NaOH) Treatment

(Following Arrighi and Hsu, 1971)

Solutions

1. Hydrochloric acid (HCl) solution (0.2 N)

2 N HCl, standardized	10 ml
Distilled water	100 ml

2. 2 × saline sodium citrate (2 × SSC) solution

Sodium chloride (NaCl)	17.5 g
Sodium citrate, 2 hydrate ($Na_3C_6H_5O_7 \cdot 2H_2O$)	8.8 g
Distilled water	1 L

Adjust pH to 7.0 using 1 N NaOH.

3. Ribonuclease (RNase) solution (100 μg/ml)

Pancreatic RNase	1 mg
2 × SSC solution	10 ml

The RNase solution should be heated for 5 to 10 minutes in a boiling water bath to destroy any deoxyribonuclease (DNase) that may be present in commercial products.

4. NaOH solution (0.07 M)

NaOH pellets	0.28 g
Distilled water	100 ml

5. Sorensen's buffer (pH 7.0)

Potassium phosphate monobasic (KH_2PO_4)	5.26 g
Sodium phosphate dibasic (Na_2HPO_4)	8.65 g
Distilled water	1 L

Instead of sodium phosphate dibasic (Na_2HPO_4), either 16.34 g of sodium phosphate dibasic, 7 hydrate ($Na_2HPO_4,7H_2O$) or 21.84 g of sodium phosphate dibasic, 12 hydrate ($Na_2HPO_4,12H_2O$) can be used to prepare the same amount of buffer.

6. Giemsa staining solution (5%)

Sorensen's buffer (pH 7.0)	47.5 ml
Gurr's Giemsa stain	2.5 ml

Procedure

1. Treat the slides (1- to 2-weeks old) in 0.2 N HCl solution for 30 minutes at room temperature. Rinse briefly in distilled water and allow to dry.
2. Place a few drops of RNase solution (100 μg/ml) on the slide, cover with a cover glass, and incubate for 1 hour at 37°C in a humid chamber. Rinse in distilled water and allow to dry.
3. Treat the slides in 0.07 M NaOH solution for 2 to 3 minutes at room temperature.

4. Rinse thoroughly in distilled water and quickly pass through a series consisting of 50 percent, 70 percent, and 100 percent alcohol. (NaOH should be removed as quickly as possible.)

5. Incubate the slides overnight in 2 × SSC (pH 7.0) in a jar preheated to 60° to 65°C.

6. Rinse the slides in two to three changes of distilled water at room temperature and allow to dry.

7. Stain the slides with 5% Giemsa staining solution in Sorensen's buffer (pH 7.0) at room temperature for 10 to 15 minutes or longer to achieve the required intensity of stain.

8. Mount the slides with permount and examine under either bright-field or phase-contrast microscope depending on the intensity of staining.

Comments

Freshly prepared slides are usually unsuitable for C-banding, because the chromosomes are too sensitive to withstand harsh treatment and lose their morphology and acquire a ghostlike hollow appearance. The slides are aged either for 1 week to 10 days at room temperature or for at least 2 to 3 days at 50° to 60°C in an oven. RNase treatment can be omitted without significant difference in the specificity or quality of C-bands. Duration of treatment in NaOH solution is critical and varies with the age of the slide as well as among different batches of slides. It is suggested that a test run of one or two slides be done for each batch. During incubation in 2 × SSC, the slides should be inspected intermittently for air bubbles that may have developed on the slide and removed by taking out and submerging the slide a few times, which otherwise produces uneven staining. Some of the treated slides may take much longer to reach the required intensity in Giemsa stain. Wet slides may be examined periodically under a microscope using low magnification optics to monitor the staining process. The chromosome preparations, especially those from bone marrow and other malignant tissues, are extremely sensitive, and in many cases treatment with NaOH may be unsuitable. These preparations may be stained according to protocol 3.21.2 which uses $Ba(OH)_2$.

Protocol 3.21.2 C-Bands Using Barium Hydroxide [Ba(OH)2] Treatment

(CBG, C-bands using barium hydroxide and Giemsa; ISCN, 1985) (Following Sumner, 1972)

Solutions

1. $Ba(OH)_2$ solution (5%)

$Ba(OH)_2$	5 g
Distilled water	100 ml

   Dissolve $Ba(OH)_2$ with magnetic stirrer for 15 to 20 minutes and filter before use. $Ba(OH)_2$ solution should be prepared fresh before use.

2. 2 × SSC
   Same as in protocol 3.21.1.

3. HCl solution (0.2 N)
   Same as in protocol 3.21.1.

4. Giemsa staining solution
   Same as in protocol 3.21.1.

Procedure

1. Treat 1- or 2-week-old slides with 0.2 N HCl for 1 hour at room temperature.

2. Rinse thoroughly in deionized water and allow to dry.

3. Treat the slides in $Ba(OH)_2$ solution (5%) for 5 to 15 minutes at 50°C in a water bath or for 30 to 40 minutes at room temperature.

4. Rinse the slides thoroughly in several changes of deionized water and pass through the alcohol series consisting of 70 percent and absolute alcohol. Allow the slides to dry.

5. Incubate the slides in $2 \times$ SSC at 60° to 65°C for 2 hours.

6. Rinse thoroughly in deionized water and allow to dry.

7. Stain with 5% Giemsa in Sorensen's buffer (pH 7.0) for 10 to 15 minutes.

Comments

$Ba(OH)_2$ is a relatively milder alkali than is NaOH, and the slides require longer treatment with $Ba(OH)_2$ to denature the chromosomes, which provides greater control over the denaturation process. Nevertheless, the extent of treatment in $Ba(OH)_2$ is critical and should be optimum to obtain quality C-bands. The duration of treatment depends on the age of the slide and the tissue of origin. $Ba(OH)_2$ treatment can also be performed at room temperature instead of 56°C (20 to 30 minutes). Treatment at low temperature preserves excellent chromosome morphology. Inadequate treatment with alkali produces poorly differentiated C-bands, whereas excessive treatment results in the loss of chromosome morphology with a hollow (ghostlike) appearance. Chromosome preparations from fibroblasts, stimulated lymphocytes, and bone marrow or other malignant tissues generally react with increasing sensitivity and therefore should be treated for shorter periods.

$Ba(OH)_2$ readily interacts with CO_2 in the atmosphere and forms barium carbonate which is sparingly soluble in water. Therefore, $Ba(OH)_2$ solution should be prepared fresh and filtered before use. The solution should be kept in a closed container during use, which will provide the required strength of $Ba(OH)_2$ and eliminate excessive deposition of barium carbonate crystals on the slides. The little deposition that may occur on slides can be rinsed off in subsequent washes. If the problem persists, a brief rinse in a dilute HCl solution will remove the crystals.

The incubation period in $2 \times$ SSC is not critical and can be adjusted to suit the working conditions. In our experience, overnight incubation gives better differentiation of C-bands, although the slides tend to pick up more background stain. (See comments in protocol 3.21.1.)

Specificity

The C-banding techniques produce selective staining of constitutive heterochromatin in chromosomes and nuclei. In humans , the C-bands are located at the centromere of all the chromosomes and the distal long arm of the Y chromosome. Chromosomes 1, 9, and 16 and the long arm of Y have significantly larger C-bands than do other chromosomes (Figure 3.2.1A and B).

Mechanism

In C-band staining, DNA is preferentially lost from the non-C-band regions of chromosomes (Comings, 1973; Pathak and Arrighi, 1973). It was suggested that two successive stages occur during the loss of DNA. Chromosomal DNA is depurinated and denatured successively during treatment in acid (HCl) and alkali [NaOH or $Ba(OH)_2$]. Denatured DNA is further broken down into smaller fragments that are lost in the solution during incubation in $2 \times$ SSC.

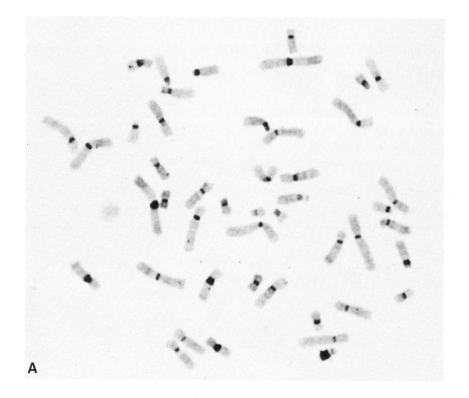

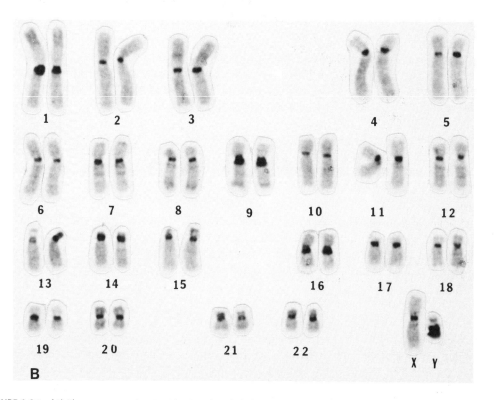

FIGURE 3.2.1. (A) Chromosomes showing the CBG-bands (C-bands by barium hydroxide using Giemsa). (See text.) (B) Karyotype of the same cell.

Although this accounts for the loss of DNA from non-C-band regions, the definite factor(s) that selectively prevent the loss of DNA from C-band regions are not yet clear. Certain non-histone proteins closely bound to C-band heterochromatin may render them selectively resistant to various treatments involved in C-banding procedures (Burkholder and Weaver, 1977; Hsieh and Brutlag, 1979).

Applications

The C-bands of respective chromosomes may vary significantly in size between homologues within an individual as well as between individuals. The variations are commonly known as C-band polymorphisms or more aptly heteromorphisms. The C-bands of virtually all chromosomes are heteromorphic (McKenzie and Lubs, 1975; Verma and Dosik, 1980). The variation in size of C-bands is continuous but not discrete. They are classified into five size categories, levels 1 to 5 (very small, small, intermediate, large, and very large) (Paris Conference, 1971, supplement to Paris Conference, 1975). The C-bands are also heteromorphic in their position in relation to the primary constriction. On the basis of their position, they can be classified into five levels, 1 to 5 (no inversion, NI; partial inversion-minor, MIN; half inversion, HI; partial inversion-major, MAJ; and complete inversion, CI) (Verma et al, 1979). Heteromorphisms are more pronounced in chromosomes 1, 9, 16 and the Y, which have larger C-bands than do the rest of the complement.

Heteromorphisms of C-bands have no proven clinical significance. The size and position of the C-band in each chromosome are characteristic and consistent in different cells and tissues. Heteromorphisms are familialy inherited; therefore, they can be used as chromosomal markers in certain cases. C-band staining is particularly useful in studying the chromosomes of unusual morphology to determine if the unusual nature is contributed to by a heteromorphic C-band. Furthermore, translocations with at least one breakpoint at the C-band region can be thoroughly investigated with C-banding to identify the precise breakpoints and to evaluate the status of the centromeric region.

3.22   NOR-BANDS

Specific chromosomal regions that form and maintain the nucleoli in interphase nuclei are called nucleolar organizer regions or nucleolus organizing regions (NORs). They consist of multiple copies of DNA sequences or genes for the larger fraction (28s) of ribosomal RNA. NORs can be differentially stained in metaphase chromosomes using either Giemsa stain (N-banding) or silver impregnation (Ag-NOR-banding) (Goodpasture and Bloom, 1975; Matsui and Sasaki, 1973). Although the N-banding method has had no major changes since its inception, the silver impregnation method has several modified versions to render it simpler and less time-consuming. The basic steps involved in both of these techniques, together with some of the modifications, are surmised.

Protocol 3.22.1   N-Banding
(NOR Staining by Giemsa Stain)

(Following Matsui and Sasaki, 1973)

Solutions

1. Trichloroacetic acid (TCA) solution (100%)

TCA	500 g
Distilled water	227 ml

(This solution is considered as 100 percent TCA.)

2. TCA working solution (5%)

TCA (100%)	5 ml
Distilled water	100 ml

3. HCl solution (0.1 N)

2.0 N HCl, standardized	5 ml
Distilled water	100 ml

4. Sorensen's buffer (pH.7.0)
   Same as in protocol 3.21.1.

5. Giemsa staining solution (10%)

Sorensen's buffer (pH 7.0)	45 ml
Gurr's Giemsa stain	5 ml

Note: The following RNase and DNase solutions are optional and are to be used as substitutes for trichloroacetic acid solution. See procedure described below.

6. RNase solution (1 mg/ml)

Pancreatic RNase	100 mg
Sorensen's buffer (pH 7.0)	100 ml

Boiling RNase solution to destroy DNase is not necessary for this protocol.

7. Sorensen's buffer (pH 6.6)

Potassium phosphate monobasic (KH_2PO_4)	8.65 g
Sodium phosphate dibasic (Na_2HPO_4)	5.10 g
Distilled water	1 L

Instead of sodium phosphate dibasic (Na_2HPO_4), either 9.65 g of sodium phosphate dibasic, 7 hydrate ($Na_2HPO_4,7H_2O$) or 12.89 g of sodium phosphate dibasic, 12 hydrate ($Na_2HPO_4,12H_2O$), can be used to prepare the same amount of buffer.

8. DNase solution (1 mg/ml)

DNase-protease free	10 mg
Sorensen's buffer (pH 6.6)	10 ml

Procedure

1. Treat the slides in TCA working solution (5%) at 85° to 90°C for 30 minutes. TCA treatment can be replaced by treatment with RNase and DNase. Treat the slide in RNase solution (1 mg/ml) at 37°C for 1 hour. Rinse in deionized water and allow to dry. Place DNase solution (1 mg/ml) on each slide, overlay with a cover glass, and incubate at 37°C in a humid chamber for 1 hour.

2. Rinse briefly in deionized water and reincubate in HCl solution (0.1 M) at 60°C for 30 minutes.

3. Rinse thoroughly in deionized water and stain in Giemsa staining solution (10%) for 60 to 90 minutes. Rinse in Sorensen's buffer (pH 7.0) and let dry. Mount the slide using permount.

4. Examine the slides using bright-field optics. If the staining intensity is too faint, phase-contrast optics may be used.

Comments
Giemsa staining of the treated slides requires a considerably long time. During staining, the slides may periodically be taken out, rinsed in buffer, and examined under low magnification to control the intensity of staining. Q-banding can be performed before N-band staining to help identify the chromosomes.

Specificity

In metaphases stained by N-banding, darkly stained regions are seen at secondary constrictions in the short arms of acrocentric chromosomes. The stained regions may appear as one or two spots of various sizes. In interphase nuclei, N-bands appear as a cluster of tiny dots within the nucleoli (Matsui and Sasaki, 1973). The N-bands coincide with the location of ribosomal cistrons (Matsui, 1974).

Mechanism

The technique used for N-banding involves extraction of both nucleic acids, DNA and RNA, and histones. It is therefore suggested that the remaining nonhistone residual proteins are the chromosomal components that are stained by Giemsa stain in the form of N-bands (Matsui and Sasaki, 1973). These proteins comprise a small portion of a total nuclear protein, and they are probably linked specifically to NORs. Furthermore, they are likely to be the structural elements of chromatin rather than the gene product (Funaki et al, 1975; Matsui, 1974).

Protocol 3.22.2 Ag-AS Technique
(NOR—Banding Using Silver Nitrate)

(Following Goodpasture and Bloom, 1975)

Solutions

1. Silver nitrate ($AgNO_3$) solution (50%)

$AgNO_3$	5 g
Distilled water	10 ml

   Store in dark vials at 2° to 5°C. This solution is fairly stable.

2. Ammoniacal silver solution

$AgNO_3$	4 g
Distilled water	5 ml
Ammonium hydroxide	5 ml

   Solution can be stored in dark vials for a few days.

3. Developing solution

Formaldehyde (37% formalin)	8 ml
Distilled water	98.5 ml

   Neutralize the formalin solution with sodium acetate crystals. Adjust to pH 5.6 with formic acid before use.

Procedure

1. Place 5 to 6 drops of $AgNO_3$ solution (50%) on a slide and cover with a cover glass.
2. Expose the slide to a flood lamp from a distance of 25 cm for 10 minutes.
3. Rinse the cover glass off the slide in distilled water.
4. Place equal amounts (4 drops of each) of ammoniacal silver solution and developing solution. Overlay the slide with a cover glass.
5. Staining intensity is monitored under a microscope using low magnification until the chromosomes are golden yellow. Quickly remove the coverslip by rinsing in distilled water and dehydrate in an alcohol series. Allow the slide to dry.
6. Mount the slide in permount and examine using either bright-field or phase-contrast microscopy.

Comments

The developing solution should be adjusted to pH 5.6 before use. Silver impregnation in Ag-AS must be controlled by microscopic observation to obtain satisfactory differential staining. This method produces a significant background of silver deposition. A number of modified versions described by other investigators have proved easier to follow and provide better control and results. QFQ can be performed before silver staining to facilitate chromosome identification.

Protocol 3.22.3   Improved Silver Staining Technique

(Following Bloom and Goodpasture, 1976)

Solutions

1. AgNO$_3$ solution (50%)
     Same as in protocol 3.22.2.

Procedure

1. Pipette 5 to 6 drops of AgNO$_3$ solution (50%) onto a slide and cover with a cover glass (24 × 50 mm).
2. Incubate the slide at 37°C in a humid chamber (to prevent drying) for 18 to 24 hours.
3. When the chromosomes are stained to the required intensity, remove the cover glass, rinse thoroughly in distilled water, and let dry.
4. Mount the slide in permount and examine.

Comments

This protocol eliminates the need to prepare several solutions and avoids critical steps. Control of staining intensity is much easier than it is with the previous method.

Protocol 3.22.4   One-Step Silver Staining Method with a Protective Colloidal Developer

(Following Howell and Black, 1980; VedBrat et al, 1980)

Solutions

1. Gelatin solution

Gelatin	2 g
Distilled water	100 ml
Formic acid	1 ml

     Dissolve gelatin by constant stirring for 10 to 15 minutes. This solution lasts 2 to 3 weeks. Discard if the solution becomes turbid.
2. AgNO$_3$ solution (50%)
     Same as in protocol 3.22.2.

Procedure

1. Place 2 drops of the gelatin solution and 4 drops of the AgNO$_3$ solution (50%) on a slide. Mix the solutions gently by tilting the slide from side to side. Cover with a cover glass.
2. Keep the slide on a slide warmer set to 70°C until the solution turns golden brown (approximately 2 minutes).

3. Remove the cover glass and rinse the slide thoroughly in distilled water. Allow the slide to dry.

4. Mount the slide with permount and examine.

Comments

The gelatin and formic acid obtained from some suppliers are unsuitable for this procedure, because they reduce the AgNO$_3$ too rapidly leading to excessive nonspecific silver grains in the background. The temperature of the slide warmer can be decreased up to 45°C, which will reduce the rate of silver staining and provide even better conditions to regulate the intensity and to examine the slide under a microscope during the process. The silver-stained slides subsequently can be banded by the trypsin-Giemsa method. As with the previous methods, for chromosome identification, the cells can be photographed for Q-bands before this procedure (Figure 3.2.2. A and B).

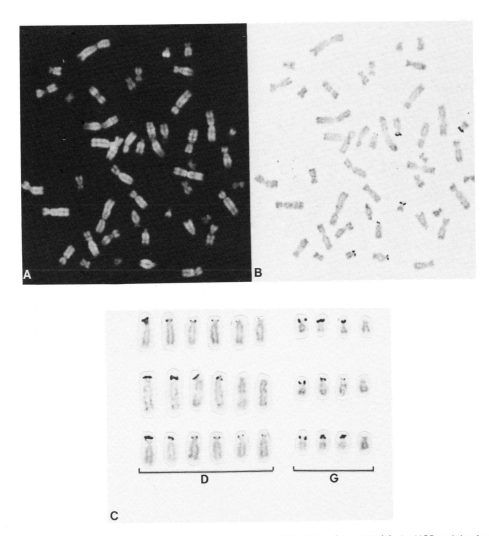

FIGURE 3.2.2. (A and B) A metaphase sequentially stained by QFQ (A) and Ag-NOR (B). Ag-NOR staining is performed by one-step silver staining method (protocol 3.22.4). Prior Q-banding helps to identify the chromosomes. (C): The D- and G-group acrocentric chromosomes from three different cells of the same individual, demonstrating the nature of silver impregnation and the extent of variation that could result as a technical artifact.

The procedure of one-step silver staining has advantages over others in terms of time required, and the quality of preparations with minimal background and uniform staining.

Specificity

Irrespective of the silver impregnation protocol used, the final patterns of silver impregnation in chromosomes are similar. The NORs are heavily stained with black silver deposition, whereas the remaining complement is lightly stained with brownish yellow color (Figure 3.2.2B). The silver staining at NORs appears as one or more dotlike structures of varying sizes and may extend beyond the NORs to the space between two acrocentrics. NORs, stained by silver, are localized in the secondary constriction regions or the so-called stalks (not the satellites) in the short arms of acrocentric chromosomes. The number of silver-stained NORs per metaphase may vary in different individuals and usually ranges between 5 and 10 (Goodpasture and Bloom, 1975; Goodpasture et al, 1976). The basic pattern of silver staining remains similar in different cells of the same individual, but it may be influenced by technical variations (Figure 3.2.2C).

Mechanism

Chromosomes have been subjected to nucleases and proteases before silver staining to examine the nature of NOR-related chromatin component that is responsible for silver staining. The silver-staining pattern of chromosomes does not change following digestion with RNase and DNase, whereas it is completely eliminated even by mild treatment with proteolytic enzymes such as trypsin or pronase. These findings reveal that the NOR-specific proteins are the target chromosomal components for silver staining (Goodpasture et al, 1976). Subsequent studies further revealed that only the *active NORs* are impregnated by silver. In other words, the silver-stained regions in metaphase chromosomes represent the *active NORs* that have participated in formation of the nucleolus in the preceding interphase stage. These observations clearly indicate that proteins associated with the transcriptional activity of ribosomal cistrons are responsible for silver impregnation (Miller DA et al, 1976; Miller OJ et al, 1976). The electron microscopic studies on chromosomes and interphase nuclei evidently showed that the silver stainable substances are protein components of the ribonucleic protein complex surrounding the NOR (Schwarzacher et al, 1978). In a nonhuman study using molecular spreads from oocytes of *Pleurodeles*, Angelier et al (1982) documented that the Ag-NOR proteins are exclusively associated with the transcriptionally active part of nucleolar genes, whereas they are absent in untranscribed spacer regions. Although the precise nature of protein(s) responsible for silver impregnation remains enigmatic, several investigators suggested them as different proteins, for example, as acidic proteins rich in sulfhydryl and disulfide groups (Buys and Osinga, 1980), proteins C_{23} and B_{23} (Busch et al, 1982), and a large subunit of the RNA polymerase (Williams et al, 1982).

Applications

Among the two methods available to selectively stain NORs, the silver staining procedures have been used more extensively than has N-banding. These differences are probably due to the ease with which each technique can be performed and the reproducibility of the results by each technique. Nevertheless, both methods can be used independently and have definite advantages because N-banding reveals the location of all the NORs including both active and inactive sites, whereas silver staining selects only the active NORs. The applications of Ag-NOR staining are so numerous that only a few major aspects can be summarized in this section.

The Ag-NOR pattern of acrocentric chromosomes is consistent within an individual and is heritable (Mikelsaar et al, 1977). These patterns, together with Q-band profiles, have been used to identify the parental origin and the stage of meiotic nondisjunction in human trisomies of acrocentric chromosomes (Verma and Dosik, 1980). With Ag-NOR staining, Miller et al (1978) showed the importance of NORs in chromosomal rearrangements. The Robertsonian translocations in mouse cell lines are more frequent between chromosomes consisting of NORs than among others. This phenomenon is likely to be true in human chromosomes as well. In humans, Ag-NOR staining has been a valuable tool in examining the status of NORs and in delineating the precise breakpoints in both Robertsonian and reciprocal translocations involving acrocentric chromosomes (Mattei et al, 1979, 1980; Mikkelsen et al, 1980). Silver staining is one of the crucial techniques in discriminating small bisatellited marker chromosomes of acrocentric origin from others that have similar size and morphology. Silver staining is used to follow the sequential changes in NOR activity during human male meiosis (Schmid et al, 1983) and to evaluate the NOR activity in malignant cells (Hubbell and Hsu, 1977).

A number of modified versions of silver staining have been adopted to stain chromosome core structure, meiotic chromosomes for light and electron microscopic studies, and histone-depleted metaphase chromosomes (Burkholder, 1983; Howell and Hsu, 1979; Pathak and Hsu, 1979).

3.23   G-11-BANDS

G-11 technique involves a modification of Giemsa staining at an alkaline pH (Bobrow et al, 1972). This method helps to selectively stain some of the heterochromatic regions in human chromosomes.

Protocol 3.23.1   G-11-Bands

(Following Bobrow et al, 1972)

Solutions

1. NaOH solution (0.1 M)

NaOH	0.4 g
Distilled water	100 ml

2. pH 11 solution

   Take 100 ml of distilled water and adjust pH to 11.0 using pH meter by slowly adding sodium hydroxide solution (0.1 M).

3. Giemsa staining solution (2%)

pH 11 solution	49 ml
Gurr's Giemsa stain	1 ml

   Prepare fresh and let stand for 5 to 10 minutes at room temperature before use.

Procedure

1. Use slides aged 1 week to 10 days. Freshly prepared slides are unsuitable because chromosomes do not retain the morphologic characteristics during the staining process.
2. Allow the staining solution to stand 5 to 10 minutes after the preparation.
3. Place the slides in the staining solution.
4. Examine the slides after 12 to 13 minutes for the extent of chromosome staining by the eosin fraction of Giemsa stain.
5. When proper intensity of eosin staining (as judged by differentiation of secondary constriction regions of chromosomes 1 and 9) is achieved, rinse the slide in distilled water and allow to dry.
6. Mount the slide with permount and examine using bright-field optics.

Comments

The G-11 technique is relatively short and simple; however, it is tricky and requires scrupulous control during the staining process. Giemsa staining solution tends to precipitate eosin rapidly at pH 11.0. Immediate use of the staining solution may cause deposition of eosin precipitate on the slides, spoiling several chromosome spreads. The rate of precipitation slows considerably after 5 to 10 minutes. The staining solution at this concentration of Giemsa stain can be used up to 1 to 1.5 hours.

The end results of differentially stained metaphases depends greatly on the age and quality of the slides. Slide preparations aged 1 week to 10 days are more suitable than are fresh slides. Chromosome preparations free of cytoplasm show distinct patterns of staining. The cytoplasmic material tends to remain blue and interferes with the deposition of eosin on the chromosomal regions. In addition, the slides prepared with lower cell density are more suitable for G-11 staining than are those with excessive material.

Quality of staining is judged by the color of the chromosomes. Properly stained metaphases show distinct eosin (reddish) stain over secondary constriction regions of chromosomes 1 and 9, whereas the arms remain light blue. The metaphases are considered understained if the entire complement remains blue with little or no reddish stain in chromosomes 1 and 9. Conversely, overstained chromosomes turn uniformly red without any differentiation; therefore, the staining process requires microscopic monitoring.

Giemsa staining solution in modified form can also be prepared in 0.3 M sodium phosphate solution (21.29 g of sodium hydrogen phosphate dibasic (Na_2HPO_4) dissolved in 500 ml of distilled water) adjusted to pH 10.4 to 11.0 using 0.1 M NaOH solution (Brito-Babapulle, 1981; Magenis et al, 1978).

Staining Pattern

By G-11 staining, metaphases show selectively stained regions in chromosomes 1, 3, 5, 7, 9, 10, 19, and the Y. The centromeric, proximal short arm and the satellite regions of acrocentric chromosomes are variably stained depending on the characteristics of the individual chromosome (Figure 3.2.3A and B). However, G-11 staining in chromosome 4 is frequent, reflecting the heteromorphic region usually located proximally in the short arm.

Mechanism

The mechanism underlying selective staining of Giemsa stain at an alkaline pH is poorly understood. There is a close correlation between the regions stained by the G-11 method and the regions consisting of satellite DNA III and DNA of other satellite groups. It can only be speculated that the regions containing one or more specific types of satellite DNAs are selectively stained by the G-11 technique (Bobrow et al, 1972; Buhler et al, 1975).

Applications

The G-11 technique has been useful in demonstrating polymorphisms and pericentric inversions in chromosome 9. The pattern of G-11 staining in chromosomes 1 and 9 has revealed the heterogeneous regions within the C-band heterochromatin (Donlon and Magenis, 1981; Magenis et al, 1978). The heterochromatic C-band region of chromosome 1 consists of two subregions that are differentiated into G-11 positive and G-11 negative regions. Similarly, the pericentric C-band of chromosome 9 shows heterogeneity by sequential G-11 and GTG-banding. G-11 staining has been used to study chromosomes in hematologic disorders, to identify chromosome 9, and also to delineate when it is involved in an abnormality (Hays et al, 1981; Morse et al, 1982). G-11 is one of the most appreciated techniques in differentiating species-specific chromosomes in human-rodent hybrid cells, in gene mapping, and in simultaneously identifying chromatid replication of human chromosomes (Alhadeff et al, 1977; Bobrow and Cross, 1974; Burgerhout, 1975; Friend et al, 1976; Westerveld et al, 1982). Furthermore, G-11 was used in human and other primate species to analyze their karyologic relation (Bobrow and Madan, 1973).

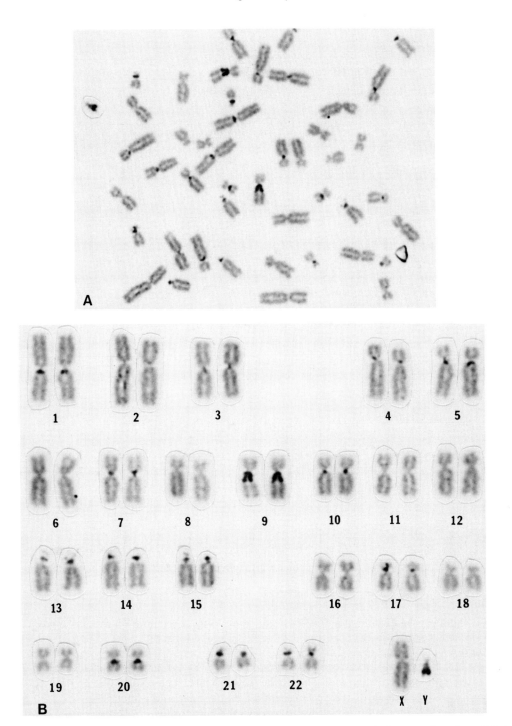

FIGURE 3.2.3. (A) Metaphase chromosome stained with Giemsa prepared in distilled water adjusted to pH 11.0, showing G-11-bands. (B) Karyotype of the same cell.

3.24   CD-BANDS

Eiberg (1974) described a new selective Giemsa staining technique that stains two dotlike bodies at the primary constriction region. These dotlike structures are considered to be the kinetochores and are therefore called centromeric dots or Cd-bands.

Protocol 3.24.1   Cd-Bands

(Following Eiberg, 1974)

1. Earle's basic salt solution (BSS) (pH 8.5–9.0)

Calcium chloride ($CaCl_2$)	0.20 g
Potassium chloride (KCl)	0.40 g
Magnesium sulfate ($MgSO_4 \cdot 7H_2O$)	0.20 g
Sodium chloride (NaCl)	6.80 g
Sodium bicarbonate ($NaHCO_3$)	2.20 g
Sodium phosphate monobasic ($NaH_2PO_4 \cdot H_2O$)	0.14 g
Distilled water	1 L

The BSS is made alkaline by allowing the CO_2 to escape from the solution and then adjusting it with 0.1 M NaOH as needed. Earle's BSS can either be prepared as just described or purchased as a ready-made solution that is commercially available.

2. Sorensen's buffer (pH 6.5)

Potassium phosphate monobasic (KH_2PO_4)	9.32 g
Sodium phosphate dibasic (Na_2HPO_4)	4.39 g
Distilled water	1 L

Instead of sodium phosphate dibasic (Na_2HPO_4), either 8.3 g of sodium phosphate dibasic, 7 hydrate ($Na_2HPO_4,7H_2O$) or 11.1 g of sodium phosphate dibasic, 12 hydrate ($Na_2HPO_4,12H_2O$) can be used to prepare the same amount of buffer.

3. Giemsa staining solution (5%)

Sorensen's buffer (pH 6.5)	47.5 ml
Gurr's Giemsa stain	2.5 ml

Prepare fresh before use.

Procedure

1. Age the slides at room temperature for 1 week to 10 days.
2. Incubate the slides in Earle's BSS preheated to 85°C for 45 minutes.
3. Rinse in deionized water and allow to dry.
4. Stain in Giemsa staining solution (5%) for 10 to 15 minutes. The optimal staining time may vary between slides, and the intensity of stain can be monitored microscopically under low magnification.
5. Rinse in deionized water and allow to dry.
6. The slides can be examined with or without mounting, depending on the objectives used.

Comments

The original method described by Eiberg (1974) includes special steps during cell fixation that utilize different ratios of methanol and acetic acid mixture. These steps can be omitted, and slides prepared according to routine procedures can be used after aging.

The basic difficulty in Cd-banding is the loss of chromosome morphology and reproducibility. Incubation time in Earle's BSS may be reduced considerably. Standardization of the procedure to obtain consistent results is essential. The metaphases can be photographed for Q-bands before Cd-banding to facilitate chromosome identification.

Staining Pattern

Each chromosome shows two spherical dotlike structures at the primary constriction region. The dots are identical within a chromosome and are of uniform size in different chromosomes of a metaphase. Prometaphase chromosomes tend to show a single narrow band instead of two independent dots. The Cd-bands are much smaller than the corresponding C-bands. The large heterochromatic regions of chromosomes 1, 9, 16, and Y do not show differential staining.

Mechanism
The mechanism of Cd-banding and the nature of the dotlike structures, Cd-bands, are the least understood and not without controversy (Evans and Ross, 1974; Roos, 1975). Few attempts have been made to resolve the true chemical nature of centromeric dots. They may represent chromosomal regions associated with spindle fibers that probably consist of specific DNA-protein complexes resistant to heat and alkali treatment (Eiberg, 1974).

Applications
Cd-banding has been used to study stable dicentric chromosomes in order to understand the nature and function of the two centromeric regions. In dicentric chromosomes, one centromere remains active or functional, whereas the other is rendered inactive or latent (Hsu et al, 1975). Cd-banding was shown to be a useful tool in discriminating active and inactive centromeres, because only active centromeres are positively stained, whereas inactive ones are not (Daniel, 1979; Maraschio et al, 1980; Nakagome et al, 1976; Romain et al, 1982). Nakagome et al (1984) suggested that chromosomes in aged women tend to lose dotlike structures (Cd-bands) or functional centromeres, which may result in aneuploid cells in mitosis and nondisjunction in meiosis.

3.3 FLUOROCHROMES AND COUNTERSTAINING TECHNIQUES

A wide variety of fluorochromes have been used to stain chromosomes and to produce characteristic fluorescence patterns. Fluorochromes (fluorescent dyes) can be broadly categorized into two different types on the basis of their affinity towards base pairs either AT or GC in DNA (Table 3.3.1). Depending on their affinity, the fluorochromes produce basic patterns similar to Q-bands or R-bands. These fluorochromes, with the exception of quinacrine, are

Table 3.3.1. DNA Ligands Used in Human Chromosome Staining and Their Base Affinity

Dye	Affinity	References
Fluorescent dyes		
DAPI	A-T	Lin et al, 1979
DIPI	A-T	Chandra and Mildner, 1979
Hoechst 33258	A-T	Latt and Wohlleb, 1975
Daunomycin	Low	Comings and Drets, 1976
7-Amino actinomycin	G-C	Gill et al, 1975
Chromomycin A_3	G-C	Jensen et al, 1977
Mithramycin	G-C	Leemann and Ruch, 1978
Olivomycin	G-C	Jorgenson et al, 1978
Quinacrine	Slightly G-C	Latt et al, 1974
Quinacrine mustard	Slightly G-C	Muller et al, 1975
Ethidum bromide	Low	Bittman, 1969
Nonfluorescent dyes		
Distamycin A	A-T	Zimmer et al, 1971
Malachite green	A-T	Muller and Gautier, 1975
Methyl green	A-T	Muller and Gautier, 1975
Crystal violet	Slightly A-T	Muller and Gautier, 1975
Actinomycin D	G-C	Muller and Crothers, 1968

seldom used for their basic fluorescence patterns. However, many of these fluorescent stains produce interesting and useful staining patterns when combined with another appropriate DNA ligand. These methods that involve the staining of chromosomes with two different chemicals, or double staining, are referred to as counterstaining. In counterstaining, the fluorochrome whose fluorescence is observed in order to examine the chromosomes is termed as a primary stain, whereas the other DNA binding agent is described as a counterstain (Schweizer, 1981). Counterstains are DNA ligands that are either nonfluorescent or fluorescent, but not in the same range as that of the primary stain. Chromosomes are generally stained with a counterstain either before or after the primary stain. There are three combinations of primary and counterstains on the basis of their affinities to DNA bases (Tables 3.3.1 and 3.3.2): (1) AT-specific primary stain and AT-specific counterstain, a combination that may produce a radically different pattern from that of the original fluorescence pattern of the primary stain, (2) AT-specific primary stain and GC-specific counterstain, which usually enhances the preexisting pattern of the primary fluorochrome, and (3) GC-specific primary stain and AT-specific counterstain, which results in enhanced R-bands of the primary fluorochrome. Some of the techniques described herein might be significant in clinical as well as nonclinical studies.

3.31   DAPI

Human chromosomes stained with 4'-6-diamidino-2-phenylindole (DAPI) show bright fluorescence at the secondary constriction regions of chromosomes 1 and 16 and a Q-band-like pattern on the remaining complement (Schweizer, 1976). If the chromosomes are exposed to a nonfluorescent counterstain distamycin A (DA), an oligopeptide antibiotic, before staining with primary stain DAPI, they exhibit an interesting pattern with certain heterochromatic regions brightly fluorescent over the rest of the complement (Schweizer et al, 1978).

Protocol 3.31.1   Distamycin A/4'-6-diamidino-2-phenylindole (DA/DAPI) Staining

(Following Schweizer et al, 1978)

Solutions

1. McIlvaine's buffer (pH 7.0)

Citric acid ($H_3C_6H_5O_7$)	0.63 g
Sodium phosphate dibasic (Na_2HPO_4)	6.19 g
Distilled water	500 ml

Instead of sodium phosphate dibasic (Na_2HPO_4), 11.69 g of sodium phosphate dibasic, 7 hydrate ($Na_2HPO_4 \cdot 7H_2O$) or 15.80 g of sodium phosphate dibasic, 12 hydrate ($Na_2HPO_4 \cdot 12H_2O$) can be used to prepare the same amount of buffer.

2. DAPI stock solution

DAPI	1 mg
Distilled water	5 ml

DAPI working solution

DAPI stock solution	50 μl
McIlvaine's buffer (pH 7.0)	15 ml

DAPI stock solution can be stored frozen in dark vials for a few months. DAPI working solution is prepared fresh before use.

Table 3.3.2. Primary Stains and Counterstains in Fluorescence Banding of Human Chromosomes*[†]

Primary Stain	Counterstain	Staining Pattern	Reference
Hoechst 33258	Distamycin A	DA/DAPI bands+	Sahar and Latt, 1980
	Netropsin	DA/DAPI bands+	Schnedle et al, 1980
	Actinomycin D	Enhanced QFH-type bands	Jorgenson et al, 1978
	7-Amino-actinomycin D	Enhanced QFH-type bands	Sahar and Latt, 1978, 1980; Latt et al, 1979; 1980
	Chromomycin A_3	Enhanced QFH-type bands	Sahar and Latt, 1978
	Echinomycin	Modified QFH-type bands	Schwizer, 1981
DAPI	Distamycin A	DA/DAPI bands	Schweizer et al, 1978
	Netropsin	DA/DAPI bands	Schnedl et al, 1980
	Pentamidine	Poor DA/DAPI bands	Schnedl et al, 1980
	Methyl green	Poor DA/DAPI bands	Schnedl et al, 1980
	Actinomycin D	Enhanced QFH-type bands	Schweizer, 1976
	Chromomycin A_3	Enhanced QFH-type bands	Schweizer 1980
	Mithramycin	Enhanced QFH-type bands	Leemann and Ruch, 1978
	Echinomycin	Modified QFH-type bands	Schnedl et al, 1980
DIPI	Netropsin	Poor DA/DAPI bands	Schnedl et al, 1980
	Pentamidine	Poor DA/DAPI bands	Schnedl et al, 1980
Chromomycin A_3	Distamycin A	Enhanced R-bands	Schweizer, 1977
	Netropsin	Enhanced R-bands	Sahar and Latt, 1980; Latt et al, 1980
	Methyl green	Enhanced R-bands	Sahar and Latt, 1978
	Actinomycin D	Enhanced R-bands	
	Crystal violet	Slightly enhanced R-bands	Schweizer, 1981
Olivomycin	Distamycin A	Enhanced R-bands	Babu, unpublished data
	Netropsin	Enhanced R-bands	Jorgenson et al, 1978
	Methyl green	Enhanced R-bands	Babu, unpublished data
7-Amino-actinomycin D	Methyl green	Enhanced R-bands	Latt et al, 1980
Mithramycin	Malachite green	Enhanced R-bands	Schnedl (unpublished data); cf. Schweizer, 1981
Coriphosphin	Methyl green	Modified R-bands	Schweizer, 1981
Quinacrine or quinacrine mustard	Actinomycin D	Enhanced Q bright polymorphic regions	Sahar and Latt, 1978, 1980; Latt et al, 1979, 1980
	Netropsin	Reduced Q fluorescence No significant effect	Lober et al, 1978 Schnedl et al, 1980
	Distamycin A	No significant effect	Schnedl et al, 1980

*Does not include the reports on other mammalian species. DAPI: 4'-6-diamidino-2-phenylindole; Hoechst 33258: (2-[2-(4-hydroxyphenyl)-6-benzimidazolyl]-6-(1-methyl-4-piperazyl)-benzimidazole).

Description of the bands:

DA/DAPI bands: Strongly fluorescent bands peracentromeric region of chromosomes 1, 9, and 16, the proximal short arm of chromosome 15, and the distal long arm region of the Y chromosome. The pericentric regions of few other chromosomes and proximal short arm regions of acrocentric chromosomes (other than 15) show slight variation in intensity.

Poor DA/DAPI bands: Bands similar to DA/DAPI but with inferior definition of DA/DAPI bands; in other words, less suppression of Q-like bands along the length of the chromosomes.

QFH-types bands: Q-like bands along the length of the chromosomes with highlighted paracentromeric regions of chromosomes 1 and 16.

Modified QFH-type bands: Distinct intercalary banding, major C-bands negative, and medium intensity on the distal long arm region of the Y chromosome.

Modified R-bands: R-bands with bright fluorescence at paracentromeric regions of some chromosomes.

[†]After Sahar and Latt (1980); Schnedl et al (1980); Schweizer (1981); and Babu (unpublished work).

3. DA solution

DA	2 mg
McIlvaine's buffer (pH 7.0)	20 ml

DA solution should be prepared fresh and kept frozen in dark vials. The solution can be used up to 24 hours.

4. Magnesium chloride solution (50 mM)

    Magnesium chloride, 6 hydrate ($MgCl_2 \cdot 6H_2O$)   100 mg
    Distilled water   10 ml

5. Mounting solution

    Glycerol   5 ml
    McIlvaine's buffer (pH 7.0)   5 ml
    Magnesium chloride solution (50 mM)   50 μl

Procedure

1. Place a few drops of DA solution on the slide and cover with a coverslip. Stain in the dark for 10 to 15 minutes at room temperature. Rinse the slide in distilled water and allow to dry.

2. Overlay the slide with a few drops of DAPI solution and cover with a coverslip. Stain in the dark for 20 to 30 minutes at room temperature. Rinse in distilled water and air dry.

3. Mount the slide with mounting solution and remove the excess by gently pressing between layers of paper. To prevent the entry of air bubbles, slides can be sealed with rubber cement.

4. Store the slides in the dark at room temperature for at least 24 to 72 hours to help stabilize the stain. If air spaces develop between the coverglass and the slide, remount and store overnight or longer.

5. The slides should be observed for fluorescence bands using neofluor objectives with the wavelength of 360 to 390 nm.

Comments

Fresh slides are preferred for DA/DAPI staining; however, older slides may also be used, but they require longer storage to stabilize the DAPI fluorescence. If air spaces develop, remount the slide with mounting medium. Remounted slides must be aged again to stabilize the fluorescence.

    Other AT-specific nonfluorescent DNA ligands such as methyl green (MG) and netropsin can be used instead of distamycin A, to obtain comparable patterns (see protocol 3.31.2) (Donlon and Magenis, 1983; Schnedl et al, 1980).

Protocol 3.31.2   Methyl Green/4'-6-diamidino-2-phenylindol (MG/DAPI) Staining

(Following Donlon and Magenis, 1983)

Solutions

1. McIlvaine's buffer (pH 4.0)

    Citric acid ($H_3C_6H_5O_7$)   2.94 g
    Sodium phosphate dibasic (Na_2HPO_4)   2.74 g
    Distilled water   500 ml

    Instead of sodium phosphate dibasic (Na_2HPO_4), 5.17 g of sodium phosphate dibasic, 7 hydrate ($Na_2HPO_4 \cdot 7H_2O$) or 6.92 g of sodium phosphate dibasic, 12 hydrate ($Na_2HPO_4 \cdot 12H_2O$) can be used to prepare the same amount of buffer.

2. McIlvaine's buffer (pH 7.0)

    Same as in protocol 3.31.1.

3. MG solution
    MG stock solution

Methyl green	1.76 g
McIlvaine's buffer (pH 4.0)	100 ml

   (The stock solution is prepared in an acidic buffer because MG keeps well at acidic pH.)

   MG staining solution

MG stock solution	1 ml
McIlvaine's buffer (pH 7.0)	50 ml

   Prepare fresh working solution before use.

4. DAPI solution
    Same as in protocol 3.31.1.

5. Mounting solution
    Same as in protocol 3.31.1.

Procedure

1. Stain the slides with MG staining solution in a jar at room temperature for 20 to 30 minutes. Rinse thoroughly in two changes of McIlvaine's buffer (pH 7.0) and allow to dry.

2. Follow steps 2 to 5 described in protocol 3.23.1 for DAPI staining and subsequent slide examination.

Comments

MG is a nonfluorescent DNA ligand with AT base-pair specificity. MG can be used as a substitute for DA to produce a banding pattern similar to that of DA/DAPI (Donlon and Magenis, 1983). The differential fluorescent bands obtained by MG/DAPI, although similar, are not as distinct as those of DA/DAPI.

Staining Pattern

Human chromosomes stained by DAPI alone show differential Q-band-like fluorescence bands along the length of the chromosomes with prominent staining at the secondary constriction regions of chromosomes 1 and 16 (Figure 3.3.1A). These bands are similar to those observed with Hoechst 33258 and are therefore termed H-bands (QFH, Q-bands by fluorescence using Hoechst 33258; ISCN, 1985). Human chromosomes stained by DA/DAPI show bright fluorescence at secondary constriction regions of chromosomes 1, 9, and 16, the proximal short arm of 15, and the distal long arm of Y. In addition, pericentric regions of some chromosomes, such as 4, 7, 10, 19, and other acrocentric chromosomes, show fluorescence of various intensities. The remaining complement is faintly fluorescent (Figure 3.3.1B and C). Chromosomes stained by MG/DAPI also exhibit a pattern similar to that of DA/DAPI, but with relatively more prominent bands along the length of the chromosomes.

Mechanism

DAPI binds to DNA with a preference towards AT base pairs. Although the fluorescence of DAPI is enhanced by both AT and GC base pairs, the enhancement is significantly greater with AT-rich than with GC-rich DNA (Lin et al, 1977). The H-bands seen in chromosomes stained by DAPI apparently reflect areas of different levels of AT base composition. DA is another DNA ligand with an affinity for an AT base pair and binds to the minor groove of DNA (Zimmer, 1975). Competitive binding between the two DNA ligands, DA and DAPI, which bind at similar but not identical sites, may lead to differential fluorescence. An alternative possibility is that the DAPI binding sites are blocked by DA in euchromatic regions, whereas at least some binding sites in heterochromatic regions are still available for DAPI (Jorgensen et al, 1978; Schweizer, 1981; Schweizer et al, 1978).

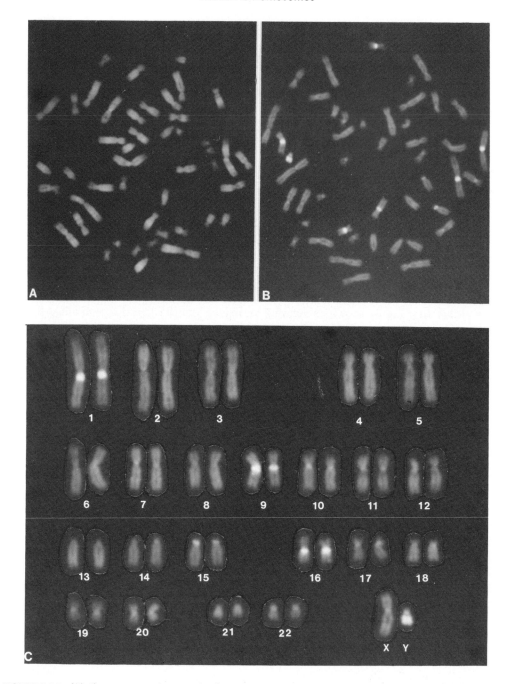

FIGURE 3.3.1. (A) Chromosomes showing the fluorescence pattern with DAPI staining. (B) Chromosomes double stained (DA/DAPI) with counterstain distamycin A (DA) and subsequently with the primary stain DAPI. (C) Karyotype of a cell stained by the DA/DAPI technique. Chromosomes 1, 9, 15, 16, and the Y show characteristic brightly fluorescent bands.

Applications

The DA/DAPI fluorescence technique is relatively simple when compared with those that produce similar patterns such as 5-methylcytocine antibodies. The specific staining pattern is useful in identifying precise breakpoints when at least one of the breakpoints is located close to selectively stained regions. DA/DAPI has been especially useful in identifying small satellited markers derived from acrocentric chromosomes. The marker chromosomes that show brilliant fluorescence on DA/DAPI staining have originated from chromosome 15 (Wisniewski and Doherty, 1985), whereas those without significant fluorescence might have originated from either chromosome 15 or others (Babu et al, 1986; Verma et al, 1985) (Figure 3.3.2A and B). Sequential DA/DAPI and C-banding have revealed heterogeneity in the heterochromatic region of chromosome 9, a distal, brightly fluorescent region and a proximal, intensely Giemsa-stained region after C-banding (Buys et al, 1981). Gosden et al (1981) showed a significant correlation between DA/DAPI fluorescence intensity and the content of satellite DNA in chromosome 9.

Triple staining of chromosomes with chromomycin A$_3$/DA/DAPI facilitates observation of the same cells by two successive banding patterns (Schweizer, 1980). (See protocol 3.34.3.)

3.32   HOECHST 33258

Chromosomes stained with Hoechst 33258 [2'-(4-hydroxyphenyl)-5-(4-methyl-1-piperazinyl)-2,5'bi-1H-benzimidazole], a bisbenzimide compound, show a Q-band-like fluorescence pattern in the complement but differ with a bright fluorescence in the secondary constriction regions of chromosomes 1 and 16 (QFH, Q-bands by fluorescence using Hoechst 33258; ISCN, 1985). Hoechst 33258 behaves similar to DAPI and gives characteristic bands when counterstained with either of two nonfluorescent antibiotics, netropsin or DA (Sahar and Latt, 1980; Schnedl et al, 1980).

Protocol 3.32.1   Distamycin A (DA)/Hoechst 33258 Staining

(Following Schnedl et al, 1980)

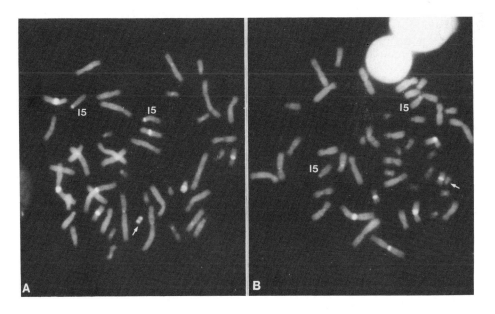

FIGURE 3.3.2. Metaphase chromosomes from two different individuals with an extra marker chromosome stained by DA/DAPI technique (A and B). The marker chromosome (arrow) in one of the individual (A) is considered to have originated from chromosome 15 because of its distinct bright fluorescence that is compatible with that of the short arm of 15 (marked). The marker chromosome (arrow) of the second individual (B) showed insignificant fluorescence intensity which could have originated from any one of the acrocentric chromosomes including polymorphic 15 (marked).

Solutions

1. DA solution (100 μg/ml)
   Same as in protocol 3.31.1.
2. Hoechst 33258 staining solution (0.5 μg/ml)

Hoechst 33258 stock solution	0.15 ml
Distilled water	45 ml

   Store in a dark bottle or coplin jar in the refrigerator. The solution can be used up to 1 month.
3. McIlvaine's buffer (pH 7.0)
   Same as in protocol 3.31.1.

Procedure

1. Flood the slides with DA solution (100 μg/ml) and stain in the dark at room temperature for 10 to 15 minutes.
2. Rinse in deionized water and allow to dry.
3. Stain the slides in Hoechst 33258 solution (0.5 μg/ml) in the dark at room temperature for 10 to 15 minutes.
4. Rinse in deionized water and allow to dry.
5. Mount in McIlvaine's buffer (pH 7.0) and blot off excess fluid.
6. Examine the slides using a wavelength of 360 to 400 nm.

Comments

Netropsin can be used instead of DA as a counterstain to obtain a similar fluorescence pattern (Sahar and Latt, 1980).

Staining Pattern

Fluorescence patterns observed by Hoechst 33258 counterstained with either DA or netropsin are similar but not as distinct as those of DA/DAPI. The secondary constriction regions of chromosomes 1, 9, and 16, the proximal short arm region of chromosome 15, and the distal long arm of Y show bright fluorescence over the rest of the complement.

Protocol 3.32.2   Hoechst 33258/Actinomycin D Staining

(Following Jorgenson et al, 1978; Sahar and Latt, 1978)

Solutions

1. Hoechst 33258 staining solution (0.5 μg/ml)
   Same as in protocol 3.32.1.
2. McIlvaine's buffer (pH 7.0)
   Same as in protocol 3.31.1.
3. McIlvaine's buffer (pH 5.5)

Citric acid ($H_3C_6H_5O_7$)	4.13 g
Sodium phosphate dibasic (Na_2HPO_4)	8.08 g
Distilled water	1 L

   Instead of sodium phosphate dibasic (Na_2HPO_4), 15.29 g of sodium phosphate dibasic, 7 hydrate ($Na_2HPO_4 \cdot 7H_2O$) or 20.43 g of sodium phosphate dibasic, 12 hydrate ($Na_2HPO_4 \cdot 12H_2O$) can be used to prepare the same amount of buffer.

4. Dilute McIlvaine's buffer (1:5)

McIlvaine's buffer (pH 5.5)	10 ml
Distilled water	50 ml

5. Actinomycin D staining solution (30 μgM)

Actinomycin D	1 mg
Dilute McIlvaine's buffer (1:5)	25 ml

Store frozen in a dark vial.

Procedure

1. Stain the slides in Hoechst 33258 staining solution (0.5 μg/ml) in the dark at room temperature for 10 to 15 minutes.
2. Rinse in deionized water and allow to dry.
3. Flood the slides with actinomycin D staining solution (30 μgM) and stain in the dark at room temperature for 15 to 20 minutes.
4. Rinse in deionized water and allow to dry.
5. Mount in McIlvaine's buffer (pH 7.0) and blot off excess fluid.
6. Examine the slides using a wavelength of 360 to 400 nm.

Comments

Other GC-specific DNA ligands, such as chromomycin A$_3$, echinomycin, and 7-amino-actinomycin D, can be used as counterstains instead of actinomycin D. However, these compounds are fluorochromes and their fluorescence interferes with that of the primary stain Hoechst 33258, which can be overcome by using a 430-nm high-pass filter.

Staining Pattern

Chromosomes stained with Hoechst 33258, when counterstained with one of the GC-specific DNA ligands, actinomycin D, show enhanced QFH-type bands (Figure 3.3.3). Similarly, chromosomes double-stained with Hoechst 33258 and chromomycin A$_3$ showed enhanced QFH-

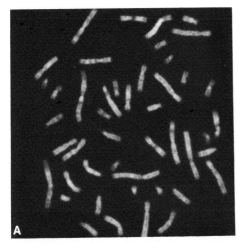 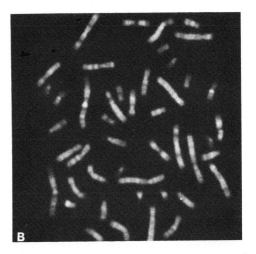

FIGURE 3.3.3. (A) Chromosomes of a male individual stained with Hoechst 33258, showing fluorescent bands. (QFH; Q-bands by fluorescence using Hoechst 33258). (B) The same cell shown is subsequently counterstained with actinomycin D and rephotographed for Hoechst 33258 fluorescence. The QFH bands, particularly those at the secondary constriction regions of chromosomes 1 and 16 and the heterochromatin in the long arm of the Y, are enhanced.

type bands produced by Hoechst 3358, but the same is not true for the R-bands of chromomycin A_3.

Mechanism

The differential fluorescence patterns of Hoechst 33258 by counterstaining with either AT-specific DNA ligand, netropsin, and distamycin A is likely to be due to competitive binding of the primary stain and counterstain as explained for DA/DAPI. (See mechanism under section 3.31.) Enhancement of the QFH-type bands of Hoechst 33258 by the actinomycins is more likely an effect of electronic energy transfer. In this case, however, the R-bands of either chromomycin A_3 or 7-amino-actinomycin D are not appreciably enhanced (Sahar and Latt, 1978).

Applications

Hoechst 33258/netropsin and Hoechst 3325/distamycin A fluorescence patterns can be used in a number of chromosome abnormalities similar to those described for DA/DAPI.

3.33   QUINACRINE

Quinacrine is one of the most frequently used fluorescent stains in cytogenetics. It is an alternative in several situations in which G-banding cannot be performed or preparations are unsuitable for G-banding. It is an ideal fluorochrome useful in sequential staining protocols to facilitate the identification of chromosomes. Furthermore, the fluorescence patterns of chromosome 3 and acrocentric groups are extremely important in studies of polymorphisms that may also serve as chromosomal markers in investigating parental origin (Olson et al, 1986). Some heteromorphisms expressed by quinacrine fluorescence can be further enhanced by subsequently counterstaining the chromosomes with actinomycin D (Sahar and Latt, 1978).

Protocol 3.33.1   Quinacrine/Actinomycin D Staining

(Following Sahar and Latt, 1978)

Solutions

1. McIlvaine's buffer (pH 5.6)
   Same as in protocol 3.21.1.

2. Quinacrine staining solution
   Same as in protocol 3.21.1.

3. McIlvaine's buffer (pH 5.5)
   Same as in protocol 3.32.2.

4. Dilute McIlvaine's buffer (1:5)

McIlvaine's buffer (pH 5.5)	10 ml
Distilled water	50 ml

5. Actinomycin D solution (30 μM)

Actinomycin D	1 mg
Dilute McIlvaine's buffer (1:5)	25 ml

   Store frozen in a dark vial.

Procedure

1. Stain the slide in quinacrine solution for 10 to 15 minutes in the dark.

2. Rinse the slide in deionized water and allow it to dry.

3. Place a few drops of actinomycin D solution (30 μM) on a slide and cover with a cover glass. Allow it to stain for 15 to 20 minutes in the dark at room temperature.

4. Rinse the slide in deionized water and allow it to dry.

5. Mount the slide in McIlvaine's buffer (pH 5.5) and blot off excess buffer.

6. Examine the slides using a wavelength between 450 and 500 nm.

Comments

Quinacrine solution is usually stored in a staining jar at 2° to 5°C. Slides can be directly stained in the jar. They can be mounted in McIlvaine's buffer (pH 5.6) for QFQ-banding and subsequently counterstained with actinomycin D to facilitate observation of the same cells by two successive fluorescence patterns. For similar results, 7-aminoactinomycin D can be used as a counterstain instead of actinomycin D.

Staining Pattern

Quinacrine fluorescence is quenched on the chromosome arms, whereas Q-polymorphisms of pericentric region of chromosome 3, short arms of acrocentric chromosomes, and the distal long arm of Y chromosome are highlighted.

Mechanism

Enhancement of fluorescence differentiation in quinacrine-stained chromosomes by counterstain is a possible result of an energy transfer mechanism between DNA-binding dyes (Sahar and Latt, 1978). When chromosomes are stained with quinacrine and counterstained with either of the GC-specific actinomycins, quinacrine possibly acts as a donor, whereas the actinomycins become acceptors of energy transfer. Quenching of quinacrine fluorescence is more pronounced in the chromosomal regions in which DNA is an admixture of base pairs AT and GC that allow the binding of donor and acceptor in close proximity to allow energy transfer. Slight predominance of a particular base pair may not significantly alter the quenching of quinacrine fluorescence. Polymorphic regions, which probably contain DNA of exclusively or long stretches of a single base pair sequence (either AT or GC), however, may enable the dyes to bind closely to allow the energy transfer and therefore remain unaffected. The polymorphic regions presumably containing AT-rich DNA remain brightly fluorescent after counterstaining. Although these mechanism offer the most plausible explanation, chromosomal proteins and differential condensation cannot be ignored.

Applications

The increased contrast of Q-polymorphisms can be used to develop chromosome markers in clinical studies, for example, to distinguish the host cells from those of the donor in bone marrow transplantation, gene mapping, and prenatal diagnosis.

3.34 CHROMOMYCIN A₃

Chromomycin A_3 is an antibiotic that has an affinity to GC base pair. Chromosomes can be stained by chromomycin A_3 for reverse fluorescent bands (R-bands). The R-bands produced by chromomycin A_3 can be significantly enhanced by counterstaining with one of the AT-specific DNA ligands. In addition, chromosomes can be triple stained by chromomycin followed by DA and DAPI which enables the observation of two successive fluorescence patterns, enhanced R-bands of chromomycin A_3 and DA/DAPI-bands, of the same cells.

Protocol 3.34.1   Chromomycin A₃ Staining

(Following Schweizer, 1976)

Solutions

1. McIlvaine's buffer (pH 7.0)
   Same as in protocol 3.31.1.

2. Magnesium chloride solution (50 mM)
   Same as in protocol 3.31.1.

3. 1/2 concentrated McIlvaine's buffer with 5-mM magnesium chloride

McIlvaine's buffer (pH 7.0)	5 ml
Magnesium chloride solution (50 mM)	0.1 ml
Distilled water	4.9 ml

4. Chromomycin A_3 solution (0.5 mg/ml)

Chromomycin A_3	1 mg
1/2 concentrated McIlvaine's buffer +	
5 mM magnesium chloride (solution 3)	2 ml

   Dissolve chromomycin A_3 in the buffer slowly (without stirring) in the refrigerator (2° to 5°C) overnight. The solution can be stored in the dark at 2° to 5°C for a few months. Older solutions tend to stain better.

5. Mounting solution
   Same as in protocol 3.31.1.

Procedure

1. Place 0.2 to 0.3 ml of chromomycin A_3 solution on a slide and cover with a coverslip. Stain at room temperature in the dark for 1 to 3 hours. Do not allow the slide to dry. Keeping the slide in a humid chamber will prevent drying.

2. Remove the coverslip gently without scratching the material. Rinse briefly in distilled water and allow to dry.

3. Mount the slide with mounting solution. Blot off excess solution by pressing between layers of filter papers. Seal the slides with rubber cement.

4. The stained slides are aged 3 to 5 days in the dark at room temperature to stabilize chromomycin fluorescence before examination. The aging period can be reduced by keeping the slides at 37°C for 1 or 2 days.

5. The slides can be examined using a wavelength of 430 to 480 nm for chromomycin bands.

Comments

Chromomycin A_3 should be dissolved slowly without stirring. Aged chromomycin solutions give better results than do freshly prepared solutions. Solutions suspected of contamination should be discarded. The freshly stained slides show rapid fading of chromomycin fluorescence. Aging of slides stabilizes the fluorescence. The slides can be aged up to 1 week before microscopic examination. The stained preparations can be stored for several months in the refrigerator. Slides aged 1 week to 10 days give the best results. However, much older slides can be stained, although the fluorescence is less intense, and they require prolonged aging before the fluorescence stabilizes. If dry spots develop between the coverslip and the slide, remount the preparations in fresh mounting solution. Remounted slides should again be aged for 2 to 3 days before examination.

Protocol 3.34.2   Chromomycin A_3/MG Staining

(Modified from Sahar and Latt, 1978)

Solutions

1. Chromomycin A$_3$ solution (0.5 mg/ml)
   Same as in protocol 3.34.1.
2. McIlvaine's buffer (pH 4.0)
   Same as in protocol 3.31.2.
3. Methyl green solution
   Same as in protocol 3.31.2.
4. Mounting solution
   Same as in protocol 3.31.1.

Procedure

1. Place a few drops of chromomycin A$_3$ solution on a slide and cover with a coverslip. Stain at room temperature in the dark for 60 minutes. The duration of staining may be expanded 2 to 3 hours if the slide is not allowed to dry.
2. Remove the coverslip, rinse the slide briefly with distilled water, and allow to dry.
3. Stain the slides in methyl green staining solution for 20 to 30 minutes at room temperature in the dark. Rinse the slide twice in deionized water and allow to dry.
4. Mount the slide with mounting solution and remove excess fluid. Age the slides to stabilize the fluorescence.
5. Examine the slides for chromomycin A$_3$ fluorescence using a wavelength of 430 to 480 nm.

Comments

Fresh slides not older than a few days to 1 week give the best results. Older slides (up to several years) may also be used; however, R-bands are usually less intense and require longer storage to stabilize the fluorescence. If fading continues to be a problem, store slides for an additional period of 1 or 2 days at 37°C or at room temperature for 1 week or longer. The stained slides may be stored in the refrigerator up to 1 year or more.

Other AT-specific DNA ligands such as distamycin A and netropsin can be used instead of methyl green following a similar protocol to enhance the chromomycin A$_3$ bands (Sahar and Latt, 1980; Schweizer, 1977).

Staining Pattern

The R-bands produced by chromomycin A$_3$ are significantly enhanced by counterstaining with methyl green and are more appreciable in chromosome identification than are those produced by chromomycin A$_3$ alone (Figure 3.3.4A, B, and C).

Protocol 3.34.3   Chromomycin A$_3$/DA/DAPI Staining

(Following Schweizer, 1980)

Solutions

1. Chromomycin A$_3$ solution
   Same as in protocol 3.34.1.
2. DAPI solutions
   Same as in protocol 3.31.1.

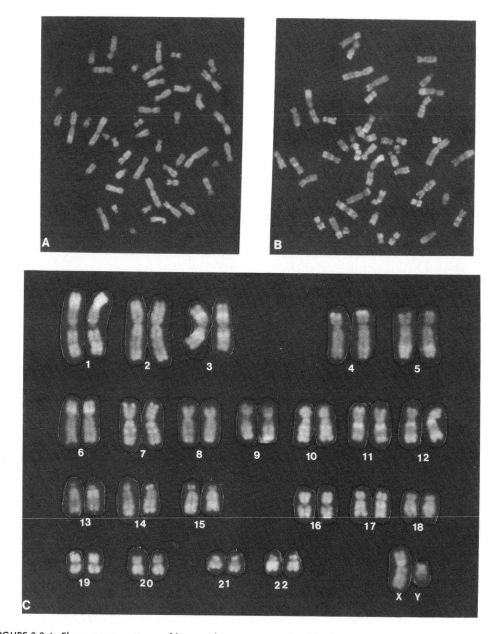

FIGURE 3.3.4. Fluorescence patterns of human chromosomes stained by chromomycin A3 (protocol 3.3.4.1) (A) and counterstained by methyl green before chromomycin A3 staining (protocol 3.3.4.2) (B). Karyotype of the cell shown above stained by methyl green/chromomycin A3 technique (C).

3. DA solution

   Same as in protocol 3.31.1.

4. Mounting medium

   Same as in protocol 3.31.1.

Procedure

1. Stain the slide with chromomycin A$_3$ solution following procedure steps 1 to 2 in previous protocol 3.34.1.

2. Subsequently, stain the slide with DA and DAPI solutions according to procedure steps 1 to 4 described in protocol 3.23.1.

3. Examine the slides for DA/DAPI bands using a wavelength of 360 to 390 nm (optimum 360 to 370 nm) and subsequently for R-bands by chromomycin A_3 using a wavelength of 430 to 480 nm (optimum 435 to 450 nm).

Comments

See comments in previous protocol 3.34.1. Photograph R-bands by chromomycin A_3 before DA/DAPI bands. Appropriate filter combinations for chromomycin A_3 and DAPI fluorescence are essential and otherwise lead to a mixture of fluorescence emitted by both the compounds and nonspecific bands. (Also see comments under protocol 3.31.1.)

Staining Pattern

Chromosomes stained by this method, when examined under appropriate optical conditions, show two successive distinct banding patterns, that is, enhanced chromomycin A_3 R-bands and selective DA/DAPI bands.

Mechanism

Chromomycin A_3 is a GC-specific DNA ligand that binds GC-rich regions of chromosomes. Thus, the R-bands produced by chromomycin A_3 reflect the GC-rich DNA in chromosomes. Although the chromomycin A_3 R-bands are enhanced by treatment with either MG, netropsin, or DA, the mechanism involved in dye combinations seems to be different. In MG counterstaining, electronic transfer of energy rather than competitive binding is indicated in increasing the contrast of chromomycin A_3 fluorescence. In double staining, which involves this mechanism, quantitative differences in the amount of dye bound to chromosome regions depicting different intensities are insignificant (Sahar and Latt, 1980). In contrast to MG, selective displacement by the AT-specific DNA ligand, either DA or netropsin, seems more likely to play an important role in enhancement of the chromomycin A_3 bands (Jorgenson et al, 1978; Schweizer, 1981).

Applications

The triple-staining method permits examination of same metaphases with two different banding patterns. It is appreciated when precise chromosome identification is required before evaluation of the same chromosomes for selective bands produced by DA/DAPI.

3.35   OLIVOMYCIN

Olivomycin is an antibiotic with an affinity for GC base pair. Lin et al (1978) showed that chromosomes stained by olivomycin exhibit a fluorescence pattern similar to that of the R-bands. The differential fluorescence of olivomycin can be further enhanced by counterstaining with one of the DNA ligands such as actinomycin D or DA.

Protocol 3.35.1   Olivomycin Staining

(Following Lin et al, 1978)

Solutions

1. Sorensen's buffer (pH 6.8)

Potassium phosphate monobasic (KH_2PO_4)	3.37 g
Sodium phosphate dibasic (Na_2HPO_4)	3.54 g
Distilled water	500 ml

Either 6.7 g of sodium phosphate dibasic, 7-hydrate ($Na_2HPO_4 \cdot 7H_2O$) or 8.96 g of sodium phosphate dibasic, 12-hydrate ($Na_2HPO_4 \cdot 12H_2O$) can be used instead of sodium phosphate dibasic (Na_2HPO_4) to prepare the same amount of buffer.

2. Olivomycin staining solution

Olivomycin	1 mg
Sorensen's buffer (pH 6.8)	1 ml

Store in the dark at 0 to 4° C.

3. Mounting solution
   Same as in protocol 3.31.1.

Procedure

1. Place a few drops of olivomycin on the slide and cover with a coverslip. Stain for 20 to 25 minutes in the dark at room temperature.

2. Rinse in two changes of Sorensen's buffer (pH 6.8) and let dry.

3. Mount in mounting solution and age the slides for 2 to 3 days at room temperature. Examine the slides at a wavelength of 450 to 470 nm.

Comments

The initial fluorescence intensity of olivomycin-stained chromosomes is impressive and adequately bright. The major problem with olivomycin in chromosome studies is rapid fading of fluorescence. Photographing the metaphases is extremely difficult even with high speed film and requires state of the art technique. We attempted to stabilize olivomycin fluorescence by mounting in a mixture of equal parts of McIlvaine's buffer (pH 7.0) and glycerol containing 0.5 mM of magnesium chloride and subsequently aging the stained slides for 2 to 3 days before examining under a microscope. This variation has been helpful in slowing down the rate of fading.

Specificity

A fluorescence banding pattern similar to that of R-bands is observed. The olivomycin–R-bands are adequate to identify individual chromosomes and heteromorphisms of acrocentric chromosomes (Figure 3.3.5A).

Protocol 3.35.2   Distamycin A (DA)/Olivomycin

(Babu, unpublished data)

Solutions

1. DA solution (100 µg/ml)
   Same as in protocol 3.31.1.

2. Sorensen's buffer (pH 7.0)
   Same as in protocol 3.21.1.

3. Olivomycin solution
   Same as in protocol 3.35.1.

4. Mounting solution
   Same as in protocol 3.31.1.

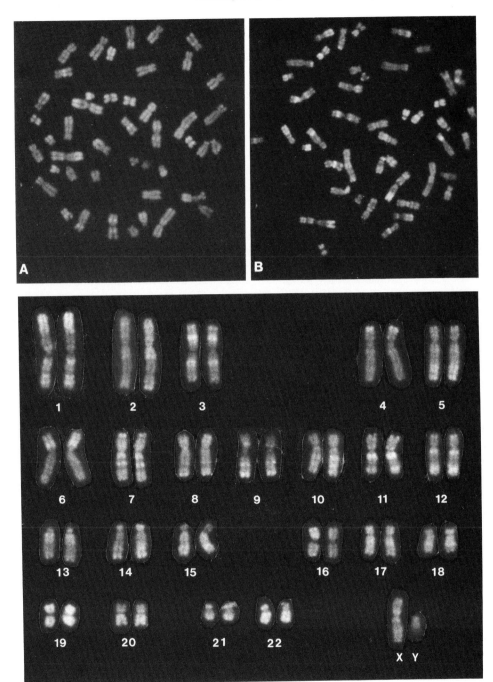

FIGURE 3.3.5. (A) Metaphase chromosomes stained by olivomycin. (B) Metaphase chromosomes counterstained by distamycin A before olivomycin staining. (C) Karyotype of a cell stained by distamycin A and olivomycin.

Procedure

1. Overlay the slides with DA solution (100 µg/ml), cover with a coverglass, and stain for 15 minutes in the dark at room temperature.

2. Rinse thoroughly in deionized water and allow to dry.

3. Subsequently follow steps 1 to 3 described for olivomycin in protocol 3.35.1.

Comments

Aging the stained slides helps to stabilize the fluorescence. Photography is performed using Kodak high speed panchromatic film Tri X. Slides can alternatively be stained by netropsin (0.5 mM) for 20 minutes in the dark at room temperature (instead of DA) before olivomycin staining to enhance R-bands (Jorgenson et al, 1978). Other DNA binding agents with AT base pair affinity, netropsin, or methyl green may be used instead of DA (Babu, unpublished data; Jorgenson et al, 1978).

Staining Pattern

Olivomycin R-bands are enhanced by prestaining with DA (Figure 3.3.5B). The enhanced R-bands are far superior for karyotypic investigations (Figure 3.3.5C).

Mechanism

Olivomycin belongs to the chromomycin family and is chemically similar to other members such as chromomycin A3 and mithramycin. Olivomycin interacts with native DNA by intercalation. The binding of olivomycin increases with an increase in the GC content of DNA. The fluorescence intensity is proportionate to the amount of olivomycin bound to DNA. The bright R-bands observed with olivomycin suggest the GC-rich regions of chromosomes (van de Sande et al, 1977). Jorgenson et al (1978) suggested that counterstaining the chromosomes with an AT-specific ligand such as netropsin, which covers three base pairs in DNA, might eliminate some potential olivomycin binding sites. This is likely to be most effective in GC sites within the AT-rich regions than in the predominantly GC-rich R-band regions. This effect can cause decreased binding of olivomycin in AT-rich regions, enhancing the contrast of chromosome fluorescence. A similar mechanism is likely in the enhancement of olivomycin R-bands by counterstaining with DA. Contrary to this, MG is likely to enhance by electron energy transfer rather than competitive binding as suggested by Sahar and Latt (1978; 1980) for chromomycin A_3/MG staining.

Application

Olivomycin staining is relatively simple and short, and it can be applied to freshly prepared slides. The use of olivomycin eliminates the necessity of pretreating the slides at high temperatures as required for other R-banding methods (van de Sande et al, 1977).

3.36   D287/170

A fluorescent dye, D287/170, was introduced as a chromosome stain by Schnedl et al (1981). This fluorochrome has an affinity to AT base pair and produces a selective and characteristic fluorescent pattern in human chromosomes. This dye does not require a counterstain for selective staining.

Protocol 3.36.1   D287/170 Staining

(Following Schnedl et al, 1981)

Solutions

1. McIlvaine's buffer (pH 7.0)
   Same as in protocol 3.31.1.
2. Diluting solution

McIlvaine's buffer	22.5 ml
Distilled water	22.5 ml
Dimethylsulfoxide (DMSO)	5.0 ml

3. D287/170 solution

D287/170 stock solution (1 mg/ml)

D287/170	10 mg
Distilled water	10 ml

Store frozen in dark vials. The solution can be used up to a few months.

D287/170 staining solution (50 μg/ml)

D287/170 stock solution (1 mg/ml)	0.25 ml
Diluting solution	4.75 ml

Store frozen in dark vials. Stain tends to precipitate. The solution can be used for 2 to 3 days.

4. Mounting solution

Sodium phosphate dibasic (Na$_2$HPO$_4$)	70 mg
Distilled water	10 ml
Glycerol	10 ml

Instead of sodium phosphate dibasic (Na$_2$HPO$_4$), 134 mg of sodium phosphate dibasic, 7 hydrate (Na$_2$HPO$_4$·7H$_2$O) or 179 mg of sodium phosphate dibasic, 12 hydrate (Na$_2$HPO$_4$·12H$_2$O) can be used to prepare the same amount of mounting solution.

Procedure

1. Place a few drops of D287/170 staining solution on a slide and cover with a cover glass. Allow the slide to stain in the dark for 20 to 30 minutes at room temperature.

2. Rinse thoroughly in deionized water and allow to dry.

3. Mount the slides in mounting medium and blot off excess mounting medium. Store the slides in black boxes overnight.

4. Examine the slides using a wavelength of 450 to 470 nm. Photography can be performed using either Kodak Tri X or Tech Pan film.

Comments

The D287/170 stock solution can be stored for a few months. The staining solution prepared in McIlvaine's buffer tends to precipitate and must be prepared fresh. The duration of staining in D287/170 staining solution can be prolonged, if needed.

Staining Pattern

The chromosomes stained by dye D287/170 show bright fluorescence of the secondary constriction region of chromosome 9, the proximal region in the short arm of chromosome 15, and the distal region in the long arm of Y. The rest of the complement is faintly fluorescent (Figure 3.3.6A and B).

Mechanism

Fluorochrome D287/170 has an affinity for AT base pair. Although little is known about the interaction of the dye with chromosomal DNA, this compound may have a greater affinity for a particular kind of repeated sequences (Schnedl et al, 1981).

Application

Fluorescent dye D287/170 is useful in investigating the heteromorphisms in chromosomes 9, 15, and the Y (Schnedl et al, 1981). Small bisatellited marker chromosomes that have originated from chromosome 15 can be identified by D287/170 staining (Figure 3.3.7A, B & C) (Babu and Verma, 1984).

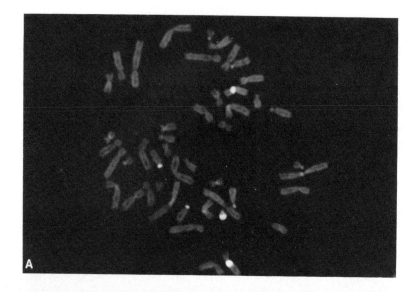

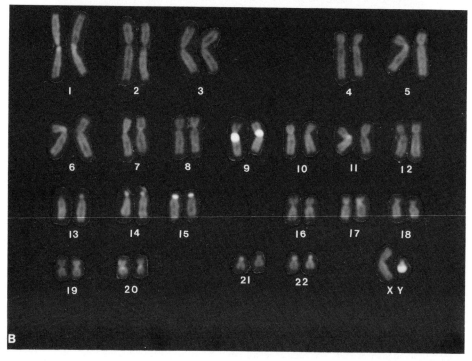

FIGURE 3.3.6. (A) A metaphase of a male stained by D287/170. (B) Karyotype of the same cell.

3.4   STAINING TECHNIQUES USING ANTIBODIES

Antibodies for specific chromosomal components are used to stain chromosomes after indirect immunofluorescence. These methods include antinucleoside antibodies generated in rabbits and antikinetochore antibodies present in the sera of patients with scleroderma of the CREST variety.

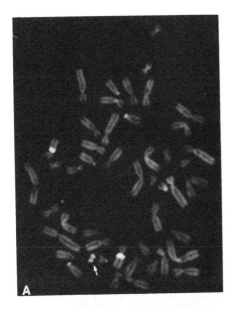
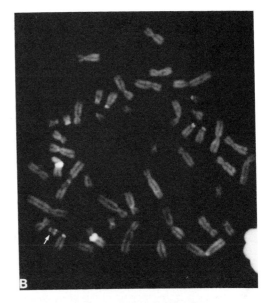
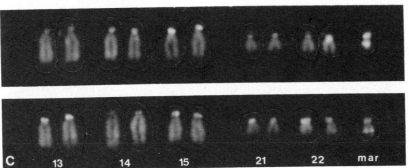

FIGURE 3.3.7. Metaphase chromosome of two individuals with an extra marker chromosome (A and B, *arrows*) stained by D287/170. The marker chromosome in A is brilliantly fluorescent and therefore considered to have originated from chromosome 15 (C, *top row*). The marker chromosome in the second individual (B) shows dull fluorescence whose identification is obscure (C, *bottom row*).

3.41   ANTI-5-METHYLCYTOCINE (ANTI-5-MEC) ANTIBODIES

Purified antibodies for specific nucleosides can be used to reveal specific classes of DNA in methanol/acetic acid fixed chromosomes by indirect immunofluorescence or immunoperoxidase staining (Miller et al, 1974). Among antibodies specific for purine and pyrimidine bases, antisera to 5-methylcytosine (5-MeC) have been used to investigate the chromosomal location of this nucleoside.

Methods of preparation and purification of rabbit antisera using protein-conjugated ribonucleosides and ribonucleotides have been described by Erlanger and Beiser (1964) and Szafran et al (1969).

Protocol 3.41.1   Anti-5-methylcytosine (anti-5-MeC) Antibody Staining

(Following Schreck et al, 1977)

Solutions

1. Rabbit antisera for 5-MeC

2. Phosphate buffered saline (PBS)

Sodium chloride (NaCl)	8.3 g
Potassium phosphate (KH_2PO_4)	6.25 ml
Sodium phosphate, dibasic (Na_2HPO_4)	35.5 ml
Distilled water	1000 ml

3. Antibody working solution

Rabbit antisera for 5-MeC (globulin fraction)	100 μl
Phosphate buffered saline	3.0 ml

4. Sheep anti-rabbit IgG-FITC (fluorescein isothiocyanate-conjugated) working solution

Sheep anti-rabbit IgG-FITC	100 μl
Phosphate buffered saline	5.0 ml

Procedure

1. Place the slide in a petri dish and cover with an adequate amount of phosphate buffered saline solution.

2. Expose to an ultraviolet source using either a germicidal lamp or black-light-blue at a distance of 15–20 cm for 18 h.

3. Rinse the slides quickly in chilled PBS and allow to dry.

4. Place 0.3 ml of antibody working solution on the slide and cover with a cover glass. Incubate at room temperature for 30 min. Remove the cover glass and rinse the slide thoroughly in several changes of PBS.

5. Place 0.5 ml of sheep anti-rabbit IgG-FITC working solution and cover with a cover glass. Incubate at room temperature for 30 minutes. Remove the cover glass and rinse thoroughly in several changes of PBS.

6. Mount the slide in PBS and examine using fluorescence microscope using a wavelength of 450 to 490 nm.

Comments

Denaturation of chromosomal DNA can be performed using alternate procedures for ultraviolet radiation. Other denaturation procedures include photooxidation using methylene blue (Schreck et al, 1973) or incubation in Sorensen's buffer (pH 6.5) at 85°C for 15 to 30 minutes (Miller et al, 1974). The slides should be rinsed in chilled PBS immediately following denaturation to maintain the denatured DNA in a single-stranded state.

Staining Pattern

Chromosomes stained by anti-5-MeC antibody staining show prominent, intensely fluorescent bands at the peracentromeric regions of chromosomes 1, 9, and 16, the proximal short arm region of chromosome 15, and the long arm of the Y chromosome. In addition, small bright bands are present at the centromeric regions of chromosomes 10, 14, 17, 22, and the Y chromosome (Miller et al, 1974). The bands in homologous pairs may vary, reflecting the heteromorphisms.

Mechanism

Antibodies for nucleosides can bind to denatured single-stranded DNA in fixed metaphase chromosomes. Antibody binding apparently requires no more than five base-pair long denatured sequences containing at least one identical base in a row (Schreck et al, 1977). The chromosomally bound antinucleoside antibodies are conjugated to fluorescein-tagged antisera for microscopic visualization.

The pattern obtained by anti-5-MeC staining of human chromosomes represents areas containing highly repetitious methylated DNA (Miller et al, 1974).

Applications

Immunofluorescence methods using antinucleoside antibodies directly demonstrate the chromosomal localization of some specific classes of DNA (Miller et al, 1974). Okamoto et al (1981) showed, that among human acrocentric chromosomes, the short arm of chromosome 15 more frequently consists of a larger collection of 5-MeC-rich DNA than those of chromosomes 13, 14, 21, and 22. This characteristic of chromosome 15 has been useful in identifying small bisatellited marker chromosomes that originate from chromosome 15 (Schreck et al, 1977). Comparative studies of heterochromatin in chimpanzees, gorillas, and humans using anti-5-MeC revealed the diversity in the size and location of 5-MeC-rich DNA, supporting the idea that satellite DNA evolves more rapidly than does the rest of the DNA (Schnedl et al, 1975).

3.42 ANTIKINETOCHORE ANTIBODIES

Sera of patients with scleroderma are known to contain autoantibodies to nucleolar and other nuclear components. Among them, patients with the CREST variety of scleroderma (>90%) have autoantibodies to kinetochores (Brenner et al, 1981). The serum of these patients is used to localize the kinetochores by indirect immunofluorescence.

Protocol 3.42.1 Antikinetochore-Antibody Staining

(Following Brenner et al, 1981, and Merry et al, 1985)

Solutions

1. Tris-HCl buffer (10 mM) (pH 7.4)

   a. Tris solution (0.2 M)

Tris	2.42 g
Distilled water	100 ml

   b. Hydrochloric acid (0.2 M)
   As obtained as an analytic reagent

   Mix:

Tris solution (0.2 M)	5.0 ml
Hydrochloric acid (HCl) (0.2 M)	4.1 ml
Distilled water	90.9 ml

   These solutions can be stored at 2° to 5°C for 6 to 8 weeks and should be discarded if turbidity develops.

2. Tris-buffered hypotonic solution

Tris-HCl buffer (pH 7.4)	100 ml	(10 mM)
Glycerol	0.29 ml	(40 mM)
Sodium chloride (NaCl)	117 mg	(20 mM)
Potassium chloride (KCl)	37.5 mg	(5 mM)
Calcium chloride, 2-hydrate ($CaCl_2 \cdot 2H_2O$)	14.7 mg	(1 mM)
Magnesium chloride, 6-hydrate ($MgCl_2 \cdot 6H_2O$)	10 mg	(0.5 mM)

   Storage conditions are the same as those for the foregoing solutions.

3. Fixative (80% ethanol)

Absolute ethanol	80 ml
Distilled water	20 ml

   Fixative should be chilled before use.

4. Propidium iodide staining solution (1 μg/ml)

Propidium iodide	0.1 mg
Distilled water	100 ml

   Store frozen in dark vials.

5. Mounting solution

Glycerol	9 ml
Dulbecco's PBS	1 ml
p-phenylenediamine	10 mg

   Store at 2° to 5°C. Solution can be used up to 4 to 6 weeks.

6. Sodium azide solution (1%)

Sodium azide	100 mg
Distilled water	10 ml

7. Antihuman IgG-FITC (fluorescein isothiocyanate conjugated) working solution

Antihuman IgG-FITC	1 ml
Dulbecco's PBS	19 ml

   Store frozen in the dark in small aliquots. Avoid repeated thawing.

Isolation and Storage of CREST Serum

Collect 15 to 20 ml of peripheral blood without heparin in a centrifuge tube from patients with the CREST variety of scleroderma. Allow the blood to stand at room temperature for 10 to 15 minutes and then centrifuge it at 3,000 rpm for 15 minutes. Collect the supernatant serum into a fresh tube without disturbing the clot. Store frozen in small aliquots or add 0.2 ml of 1 percent sodium azide solution for each 10-ml volume of serum to give a final concentration of 0.02 percent sodium azide, and store at 2° to 4°C.

Procedure

1. Treat the actively growing cell cultures with colcemid to accumulate mitosis. (Add 0.02 ml of colcemid (10 μg/ml) to 10 ml of lymphocyte culture for 1 hour or 0.25 ml of colcemid to 5 ml of fibroblast culture for 2.5 hours.)

2. Collect the cells in the form of a pellet by direct centrifugation in the case of lymphocyte cultures or by loosening the cells using mild trypsinization and centrifugation at 800 rpm for 8 minutes in the case of fibroblast cultures. If cultures are grown as monolayers on coverslips, skip steps 2 through 4.

3. Discard the supernatant and suspend the cells in 10 ml of tris-buffered hypotonic solution. Incubate at 4°C for 15 minutes.

4. Place 0.2 to 0.3 ml of cell suspension in each cytocentrifuge well (cytobucket) and centrifuge the cells onto a clean glass slide at 1,000 rpm for 10 minutes. Discard the supernatant and gently remove the slide from the cytobuckets.

5. Fix the cells by immersing the slides (or coverslips with monolayers) in 80 percent ethanol chilled to −20°C for 30 minutes. Remove the slide and dry by blowing gently.

6. Place the slides in Dulbecco's PBS for 5 minutes at room temperature.

7. Remove the slides, drain excess PBS (do not dry), and overlay with an adequate amount of antikinetochore antiserum (about 10 to 50 μl, depending on the number of locations

and the amount of area covered by the cells on slides or coverslips) and incubate at 37°C for 30 minutes in a humid chamber).

8. Rinse in two changes of PBS for 15 minutes each.

9. Overlay the cells with antihuman IgG-FITC working solution (about 10 to 50 μl) and incubate at 37°C for 45 minutes in the dark. Place them in a humid chamber to avoid drying.

10. Rinse in two changes of PBS for 15 minutes each in the dark.

11. Stain either with propidium iodide staining solution (1 μg/ml) for 1 minute or with DAPI staining solution (0.8 μg/ml) (protocol 2.31.1) for 15 minutes and rinse thoroughly in PBS.

12. Mount in mounting solution and blot off excess fluid by gently pressing the slide between layers of absorbent paper.

13. Examine the slides using a wavelength of 420 to 440 nm for both chromosome and kinetochore fluorescence patterns if counterstained with propidium iodide. If stained with DAPI, locate the dividing cells using a wavelength of 360 to 380 nm for DAPI fluorescence and examine subsequently, using a wavelength of 420 to 440 nm for fluorescein fluorescence.

Comments

The mitotic index in preparations could be increased using a synchronization technique for lymphocyte cultures by methotrexate block (protocol 4.11.1, steps 1 to 9) or using selective detachment of dividing cells in fibroblast cultures. Hypotonic solutions with either phosphate buffer, low concentration of salts, or excessive monovalent cations produce inferior results. Fixation before centrifuging the cells onto glass slides is unnecessary and may interfere with chromosome spreading (Stenman et al, 1975). Counterstaining with propidium iodide facilitates visualization of whole chromosomes with red fluorescence without interfering with kinetochore fluorescence, but it does not help to identify the chromosomes. Conversely, preparations stained with DAPI show QFH-type DAPI bands, facilitate chromosome identification, but cannot be seen simultaneously with kinetochore fluorescence. These preparations require two successive steps, one using a filter combination for DAPI fluorescence and another filter combination for fluorescein fluorescence to facilitate microscopic examination and/or photography.

Staining Pattern

Antikinetochore-antibody staining shows specific fluorescence at centromeres in metaphase chromosomes and interphase nuclei of human and several mammalian cells (Brenner et al, 1987; Brinkley et al, 1984; Moroi et al, 1978). In chromosomes the pattern appears as two small fluorescent spheres at the centromeric region (Figure 3.4.1), whereas in interphase nuclei they appear either as single or paired spheres apparently depending on the stage of the nucleus in the mitotic cycle. By immunoelectron microscopy, the antiserum is observed to be specific to the kinetochore component of the centromere (Brenner et al, 1987).

Mechanism

Patients with the CREST variety of scleroderma have autoantibodies that are specific to the kinetochore portion of the centromere. When cell preparations are treated with CREST sera, the antibodies bind to kinetochores. The kinetochore-antibody complex is visualized for microscopic examination by coupling it with fluorescein isothiocyanate conjugated antihuman IgG.

Initial studies of Moroi et al (1980) revealed that the centromeric antigen that binds to antibodies is a protein or polypeptide tightly bound to DNA. Subsequent investigations indicated

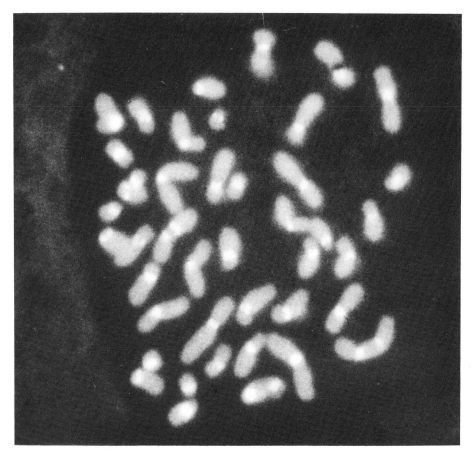

FIGURE 3.4.1. Metaphase chromosomes stained by indirect immunofluorescence technique using CREST serum. (Courtesy of Dr. S. Pathrak.)

that there is a family of antigenic centromere proteins designated as CENtromere Protein-A (CENP-A), CENP-B, and CENP-C. Sera from different patients contain different mixtures of the antibody species (Earnshaw and Rothfield, 1985).

Applications

Immunofluorescence is useful in localizing the macromolecules in cells, nucleus, and chromosomes. Kinetochores can be visualized in the light microscope using antikinetochore antibodies. Antibody staining enables us to locate the kinetochore or centromere and to follow their course and position throughout the entire mitotic cycle (Brenner et al, 1981). In-depth analysis of the chromosomal constituent that binds to antibodies would provide additional information on the chemical makeup of the kinetochore structures.

In addition to studying the basic chromosome structure, antikinetochore-antibody staining has several other applications. A comparative study of kinetochore structure using kinetochore antibodies in two taxonomically related species of deer, Indian and Chinese muntjacs, revealed an interesting phenomenon of compound kinetochores that might have originated in the process of evolution from a nonrandom aggregation of unit kinetochores (Brinkley et al, 1984). Merry et al (1985) used kinetochore antibodies as probes to examine the status of the kinetochore structures in stable dicentric chromosomes (functionally monocentric) that have originated from translocations. In their study, a dicentric translocation product was noted to have an intensely fluorescent kinetochore, which represents the functional centromere, and a less intensely fluorescent inactive kinetochore. Additional studies in this direction are likely to provide molecular and/or structural differences between active and inactive kinetochores that might unravel the mechanism of the inactivation process. Furthermore, antikinetochore-antibody fluorescence staining was used for fluorometric analysis of the size of kinetochores in human complement. It was observed that the Y chromosome has consistently smaller kinetochores than do the others which may be related to frequent aneuploidy involving the loss of the Y chromosome (Cherry and Johnston, 1987).

3.5 RESTRICTION ENDONUCLEASE/GIEMSA BANDING

The metaphase chromosomes prepared by methanol acetic acid fixation and air drying technique are susceptible to DNases (Alfi et al, 1973). Jones (1977) described that the human chromosomes digested by *Hae*III revealed G-like banding. Subsequently, Lima-de-Faria and her group (1980), using endonucleases *Hae*III and *Eco*RI to digest the chromosomes of Indian muntjac, reported that the chromosomes treated with *Hae*III showed a "hairy" appearance. The study by Miller and her group (1983), using a number of restriction endonucleases, revealed that the human chromosomes treated with some enzymes with four or five base-pair recognition sequences and stained by Giemsa showed a distinct differential staining pattern (Table 3.5.1). The staining pattern produced by each enzyme is very characteristic and reproducible (Table 3.5.2), providing the opportunity to use these enzymes in clinical studies and in understanding the nature and characteristics of the C-heterochromatic portion of the genome.

3.51 *ALU*I/GIEMSA

Protocol 3.51 *Alu*I/Giemsa Banding

(Following Miller et al, 1983, and Mezzanotte et al, 1983a)

Solutions

1. Tris-HCl buffer (pH 7.6)

   Tris solution (0.2 M)

Tris	2.42 g
Distilled water	100 ml

   Hydrochloric acid solution (0.2 N)

Hydrochloric acid standardized solution (2N)	10 ml
Distilled water	90 ml

   Stock buffer (0.1 M)

Tris solution (0.2 M)	50.0 ml
Hydrochloric acid solution (0.2 N)	29.6 ml
Distilled water	20.4 ml

   Working buffer (6 mM, pH 7.6)

Stock buffer (0.1 M)	6 ml
Distilled water	94 ml

   The concentrations of Tris and HCl are adjusted to give the proper pH at a temperature of 37°C.

Table 3.5.1. Effect of Some Restriction Endonucleases on Human Chromosomes*

Endonuclease	Recognition Sequence	Giemsa Staining Pattern
*Alu*I	5′..AG$\downarrow$CT..3′	Modified C-bands
*Dde*I	5′..C$\downarrow$TNAG..3′	Modified C-bands
*Hae*III	5′..GG$\downarrow$CC..3′	Major C-bands and G-bands
*Hinf*I	5′..G$\downarrow$ANTC..3′	Uniformly reduced staining including major C-bands except the heteromorphic regions in chromosomes 3, 4, and p arms of acrocentrics
*Mbo*I	5′..$\downarrow$GATC..3′	Modified C-bands
*Rsa*I	5′..GT$\downarrow$AC..3′	Modified C-bands

*After Miller et al, 1983; Bianchi et al, 1985a; Mezzanotte et al, 1983a, 1985; Babu and Verma, 1986a, 1987c.

Table 3.5.2. Banding Patterns Induced by Endonucleases in Human Chromosomes*¶

Chromosome No.	*Alu*I AG′CT	*Dde*I C′TNAG	*Hae*III GG′CC	*Hinf*I G′ANTC	*Mbo*I ′GATC	*Rsa*I GT′AC
1	+	+	+	−	−	+
2	−	−	−	−	+[†]	+
3	+[‡]	+	−	+[‡]	+	−
4	+[‡]	−	−	+[‡]	+[†]	−
5	+	−	−	−	+	+[†]
6	+[†]	−	−	−	+[†]	−
7	+	−	−	−	+	+
8	−	−	−	−	−[§]	−
9	+	+	+	−	+	+
10	+[§]	−	−	−	+	+[†]
11	−	−	−	−	−	+
12	−	−	−	−	+	+[§]
13	+[†]	+	+[†]	−	+	+[†]
14	+[†]	+	+[†]	−	+	+[†]
15	+	+	+	−	+	+
16	+	+	+	−	+	+
17	+	−	−	−	+[†]	+
18	+[‡]	−	−	−	+	−
19	+[§]	−	−	−	+[†]	+
20	+[†]	−	−	−	+	+
21	+[†]	+	+[†]	−	+	+[†]
22	+[†]	−	+[†]	−	+	+[†]
X	−	−	−	−	−	−[†]
Y	+	+	+	−	+	−[†]

*+ = Endonuclease-resistant band stained darkly by Giemsa; − = no band; [†] = bands vary in their intensity between homologues; [‡] = present only in heteromorphic chromosomes; [§] = disparity in the literature.

¶Following Babu and Verma, 1986; Bianchi et al, 1985.

2. Diluting solution

Working buffer (6 mM, pH 7.6)	100 ml	
Sodium chloride (NaCl)	292 mg	(50 mM)
Magnesium chloride (MgCl$_2$)	120 mg	(6 mM)
Bovine serum albumin	10 mg	(100 μg/ml)
2-Mercaptoethanol	41 μl	(6 mM)

Solutions 1 and 2 are stored at 2° to 5°C for 6 to 8 weeks and should be discarded if turbidity develops.

3. Incubating solution

Diluting solution is used to dilute the stock enzyme solution supplied by the manufacturer to obtain the required concentration of endonuclease *Alu*I. (Suggested final concentration for *Alu*I is 200 u/ml.)

Store frozen as small aliquots. Avoid repeated thawing.

Procedure

1. Place 40 μl of incubating solution on a slide, overlay with a 22 × 22-mm coverslip, and incubate in a prewarmed humid chamber. Incubation time ranges from 3 to 6 hours depending on the batch of enzyme and the age of the slide.

2. At the end of the incubation period, rinse the slides thoroughly in deionized water and allow to dry.

3. Stain the slides with 2 percent Giemsa in Sorensen's buffer (pH 7.0) at room temperature for 10 to 12 minutes. It may be necessary to monitor staining intensity under a low power microscope during the staining process.

4. Mount the slides in permount and examine using either bright-field optics or phase-contrast optics depending on the staining intensity.

Comments

See general comments at the end of this section.

Staining Pattern

The metaphases stained by *Alu*I/Giemsa technique show an overall reduction in staining from all chromosomal regions except some of the C-band regions (Figure 3.5.1A). The centromeric C-band regions in chromosomes 1, 5, 7, 9, 10, 16, and 19 and the distal heterochromatic region in the long arm of chromosome Y show consistent dark bands. The short arms of acrocentric chromosomes show various intensities and heteromorphic forms (Figure 3.5.1B). Some, but not all, chromosomes: 3, 4, 6, 17, 18, and 20 show darkly stained paracentromeric bands at the primary constriction region which represent a heteromorphic subset of C-bands (Table 3.5.2).

3.52   DDEI/GIEMSA

Protocol 3.52   *Dde*I/Giemsa Banding

(Following Miller et al, 1983)

Solutions

1. Tris-HCl buffer (pH 7.5)

Tris solution (0.2 M)
  Same as in protocol 3.51.

Hydrochloric acid solution (0.2 N)
  Same as in protocol 3.51.

Stock buffer (0.1 M)

Tris solution (0.2 M)	50.0 ml
Hydrochloric acid solution (0.2 N)	32.5 ml
Distilled water	17.5 ml

Working buffer (6 mM, pH 7.5)

Stock buffer (0.1 M)	6 ml
Distilled water	94 ml

The concentrations of Tris and HCl are adjusted to give the proper pH at a temperature of 37°C.

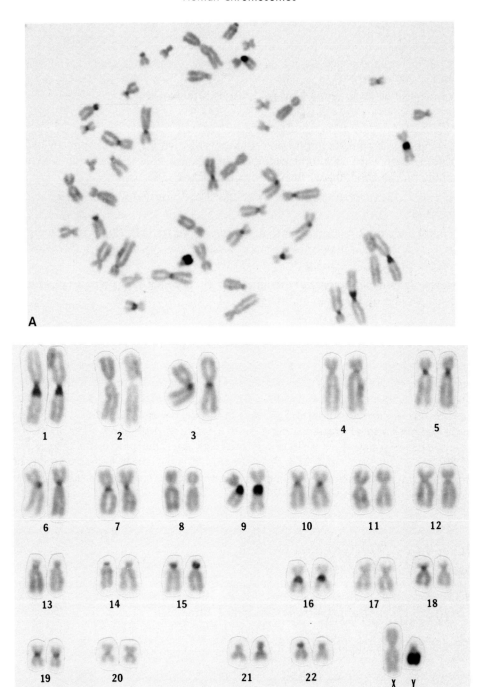

FIGURE 3.5.1. (A) A male metaphase stained by the endonuclease AluI/Giemsa technique (protocol 3.51). Digested with 200 U of AluI per milliliter of incubation medium for 5 hours at 37°C, stained in 2% Giemsa at pH 7.0, and photographed under phase-contrast microscopy using technical pan film. (B) Karyotype of the metaphase shown in A.

2. Diluting solution

Working buffer (6 mM, pH 7.5)	100 ml	
Sodium chloride (NaCl)	875 mg	(150 mM)
Magnesium chloride (MgCl$_2$)	120 mg	(6 mM)

| Bovine serum albumin | 10 mg | (100 μg/ml) |
| 2-Mercaptoethanol | 41 μl | (6 mM) |

3. Incubating solution

Diluting solution is used to dilute the stock enzyme solution supplied by the manufacturer to obtain the required concentration of endonuclease *Dde*III. (Suggested concentration for *Dde*I is 200 U/ml.)

Storage conditions are those described in the previous protocol.

Procedure

Same as in protocol 3.51.

Comments

See general comments at the end of this section.

Staining Pattern

The chromosomes digested with endonuclease *Dde*I show significant reduction in Giemsa staining except for some of the C-bands regions (Figure 3.5.2A). Dark Giemsa bands are seen near the centromeres of chromosomes 1, 3, 9, and 16 and in the long arm of the Y chromosome (Table 3.5.2). The acrocentric chromosomes show a wide range of staining patterns (Figure 3.5.2B).

3.53   HAEIII/GIEMSA

Protocol 3.53   *Hae*III/Giemsa Banding

(Following Miller et al, 1983)

Solutions

1. Tris-HCl buffer (pH 7.4)

   Tris solution
   Same as in protocol 3.51.

   Hydrochloric acid solution (0.2 N)
   Same as in protocol 3.51.

   Stock buffer (0.1 M)

Tris solution	50.0 ml
Hydrochloric acid solution (0.2 N)	35.4 ml
Distilled water	14.6 ml

   Working buffer (6 mM, pH 7.4)

   | Stock buffer | 6 ml |
   | Distilled water | 94 ml |

   The concentrations of Tris and HCl are adjusted to give the proper pH at a temperature of 37°C.

2. Diluting solution

Working buffer (6 mM, pH 7.4)	100 ml	
Sodium chloride (NaCl)	292 mg	(50 mM)
Magnesium chloride (MgCl$_2$)	120 mg	(6 mM)
Bovine serum albumin	10 mg	(100 μg/ml)
2-Mercaptoethanol	41 μl	(6 mM)

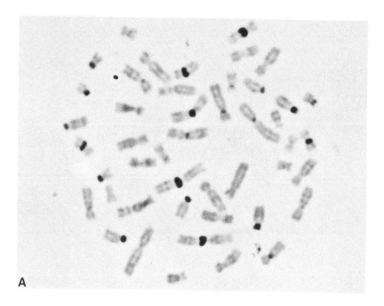

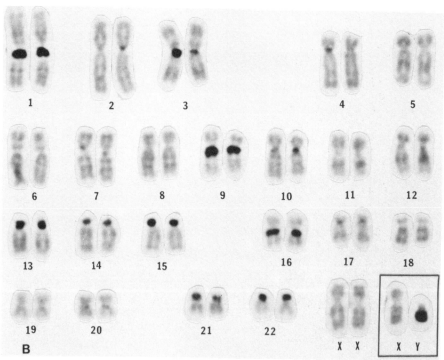

FIGURE 3.5.2. (A) A female metaphase stained by the endonuclease *Dde*I/Giemsa technique (protocol 3.5.2). Treated with 250 U of *Dde*I per milliliter in incubation medium for 4 hours, stained for 10 minutes in 2% Giemsa at pH 7.0, and photographed using phase-contrast optics and technical pan film. (B) Karyotype of the metaphase shown in A. The inset shows the sex chromosome from a male cell stained by the same protocol 3.52.

3. Incubating solution

Diluting solution is used to dilute the stock enzyme solution supplied by the manufacturer to obtain the required concentration of endonuclease *Hae*III. (Suggested concentration for *Hae*III is 400 U/ml.)

For storage conditions see protocol 3.51.

Procedure
Same as that in protocol 3.51.

Comments
See general comments at the end of this section.

Staining Pattern
Unlike other endonucleases, *Hae*III induces differential staining along the length of the chromosomes similar to the G-bands and prominent dark bands at the centromeric regions of chromosomes 1, 9, and 16 and in the distal long arm of chromosome Y. The short arms of acrocentric chromosomes show various staining intensities (Figure 3.5.3A and B).

3.54 HINFI/GIEMSA

Protocol 3.54 *Hinf* I/Giemsa Banding

(Following Miller et al, 1983)

Solutions

1. Tris-HCl buffer (pH 7.4)

   Tris solution
   Same as that in protocol 3.51.

   Hydrochloric acid solution (0.2 N)
   Same as that in protocol 3.51.

   Stock buffer (0.1 M)

Tris solution (0.2 M)	50.0 ml
Hydrochloric acid solution (0.2 N)	35.4 ml
Distilled water	14.6 ml

   Working buffer (6 mM, pH 7.4)

Stock buffer (0.1 M)	6 ml
Distilled water	94 ml

   The concentrations of Tris and HCl are adjusted to give the proper pH at a temperature of 37°C. *Note*: The buffer solution for *Hinf* I is similar to that of *Hae*III.

2. Diluting solution

Working buffer (6 mM, pH 7.4)	100 ml	
Sodium chloride (NaCl)	585 mg	(100 mM)
Magnesium chloride (MgCl$_2$)	120 mg	(6 mM)
Bovine serum albumin	10 mg	(100 μg/ml)
2-Mercaptoethanol	41 μl	(6 mM)

   Note: The concentrations of sodium chloride (NaCl) for *Hinf* I and *Hae*III are different.

3. Incubating solution

   Diluting solution is used to dilute the stock enzyme solution supplied by the manufacturer to obtain the required concentration of endonuclease *Hinf* I. (Suggested concentration for *Hinf* I is 400 U/ml.)
   For storage conditions see protocol 3.51.

Procedure
Same as that in protocol 3.51.

Comments
See general comments at the end of this section.

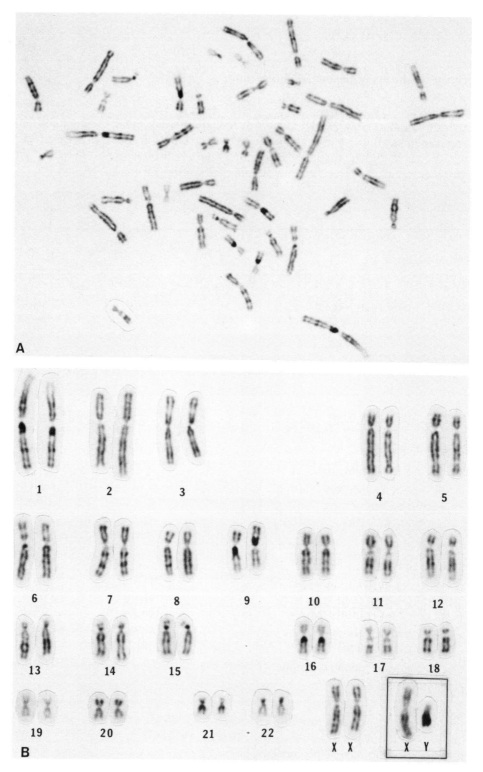

FIGURE 3.5.3. (A) A female metaphase stained by the endonuclease HaeIII/Giemsa technique (protocol 3.53). Digested with 400 U of HaeIII per milliliter of incubation medium for 6 hours at 37°C, stained in 2% Giemsa at pH 7.0, and photographed under phase-contrast microscopy using technical pan film. (B) Karyotype of the same metaphase shown in A. The inset shows the X and the Y from a male metaphase stained by the same protocol 3.53.

Staining Pattern

The endonuclease *Hinf*1 treatment extensively reduces chromosome staining from almost all regions (Table 3.5.2). The major C-bands in chromosomes 1, 9, 16, and the Y are either uniformly or negatively stained. Some chromosomes 3 and 4 show paracentromeric dark bands which apparently reflect the heteromorphic segments of corresponding C-bands (Figure 3.5.4A). The satellite regions and proximal short arm regions of some of the acrocentric chromosomes show prominent bands (Figure 3.5.4B).

3.55   MBOI/GIEMSA

Protocol 3.55   *Mbo*I/Giemsa Banding

(Following Miller et al, 1983)

Solutions

1. Tris-HCl buffer (pH 7.4)

   Tris solution (0.2 M)
   Same as in protocol 3.51.

   Hydrochloric acid solution (0.2 N)
   Same as in protocol 3.51.

   Stock buffer (0.1 M)

Tris solution (0.2 M)	50.0 ml
Hydrochloric acid solution (0.2 N)	35.4 ml
Distilled water	14.6 ml

   Working buffer (10 mM, pH 7.4)

Stock buffer (0.1 M)	10 ml
Distilled water	90 ml

   The concentrations of Tris and HCl are adjusted to give the proper pH at a temperature of 37°C.
   Note: The molarity of working buffer for *Mbo*I differs from that of *Hae*III and *Hinf*I.

2. Diluting solution

Working buffer (10 mM, pH 7.4)	100 ml	
Sodium chloride (NaCl)	585 mg	(100 mM)
Magnesium chloride ($MgCl_2$)	200 mg	(10 mM)
Bovine serum albumin	10 mg	(100 μg/ml)
Dithiothreitol	16 mg	(1 mM)

3. Incubating solution

   Diluting solution is used to dilute the stock enzyme solution supplied by the manufacturer to obtain the required concentration of endonuclease *Mbo*I. (Suggested concentration for *Mbo*I is 400 U/ml.)
   Storage conditions are as in protocol 3.51.

Procedure

Same as in protocol 3.51.

Comments

See general comments at the end of this section.

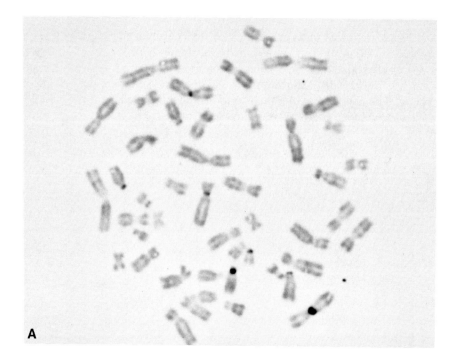

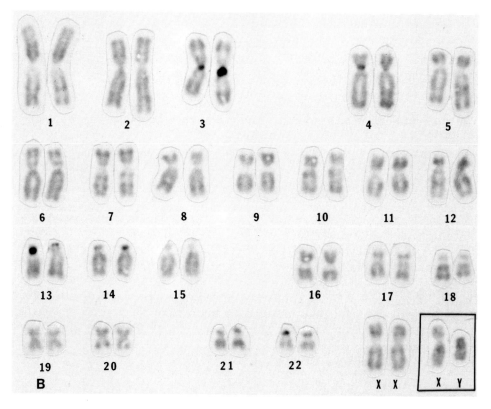

FIGURE 3.5.4. (A) A female metaphase stained by the endonuclease *Hinf* I/Giemsa technique (protocol 3.54). Treated with 400 U of *Hinf* I in incubation medium for 6 hours at 37°C, stained with Giemsa at pH 7.0, and photographed using phase-contrast optics and technical pan film. (B) Karyotype of the female metaphase shown in A. The inset shows the sex chromosomes from a male metaphase stained by the same protocol 3.54.

Staining Pattern

Metaphases treated with endonuclease *Mbo*I show an overall reduction in staining except for some of the C-band regions (Table 3.5.2). Chromosomes 3, 5, 7, 10, 12, 17, 18, and 20 show darkly stained centromeric bands (Figure 3.5.5A). The distal heterochromatic region in the long arm of the Y chromosome is darkly stained. The short arms of acrocentric chromosomes generally show dark staining. Chromosomes 2, 4, 6, and 19 show small bands at the primary constriction region (Figure 3.5.5.B). Residual bands are usually adequate for chromosome identification.

3.56   RSAI/GIEMSA

Protocol 3.56   *Rsa*I/Giemsa Banding

(Following Miller et al, 1983)

Solutions

1. Tris-HCl buffer (pH 8.0)

    Tris solution (0.2 M)
       Same as in protocol 3.51.

    Hydrochloric acid solution (0.2 N)
       Same as in protocol 3.51.

    Stock buffer (0.1 M)

Tris solution (0.2)	50.0 ml
Hydrochloric acid solution (0.2 N)	30.8 ml
Distilled water	19.2 ml

    Working buffer (6 mM, pH 8.0)

Stock buffer	6 ml
Distilled water	94 ml

    The concentrations of Tris and HCl are adjusted to give the proper pH at a temperature of 37°C.

2. Diluting solution

Working buffer (6 mM, pH 8.0)	100 ml	
Sodium chloride (NaCl)	292 mg	(50 mM)
Magnesium chloride (MgCl$_2$)	240 mg	(12 mM)
Bovine serum albumin	10 mg	(100 μg/ml)
2-Mercaptoethanol	41 μl	(6 mM)

3. Incubating solution

    Diluting solution is used to dilute the stock enzyme solution supplied by the manufacturer to obtain the required concentration of endonuclease *Rsa*I. (Suggested concentration for *Rsa*I is 400 U/ml.)

    Storage conditions are the same as in protocol 3.51.

Procedure

Same as in protocol 3.51.

Comments

See general comments at the end of this section.

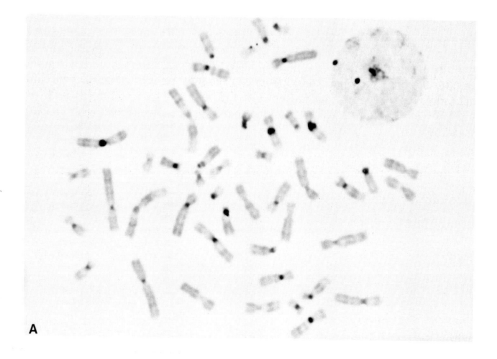

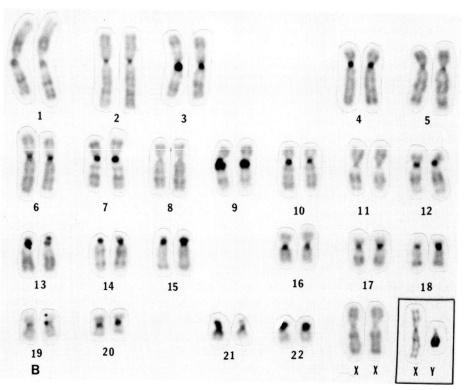

FIGURE 3.5.5. (A) A metaphase from a female amniocyte. Cell culture stained by the endonuclease MboI/Giemsa technique (protocol 3.55). Digested with 250 U of MboI in incubation medium for 5 hours at 37°C, stained in 2% Giemsa at pH 7.0, and photographed under phase-contrast microscopy using technical pan film. (B) Karyotype of the metaphase shown in A. The inset shows the X and Y chromosomes from a male cell stained by the same protocol 3.55.

Staining Pattern

The endonuclease *Rsa*I/Giemsa technique produces a modified C-band pattern with reduction in Giemsa staining from chromosomes in general (Figure 3.5.6B and C). The residual bands are meager after this procedure and are not quite similar to the G-band pattern. Therefore, the identification of chromosomes is difficult and erroneous. Prestaining the chromosomes for Q-bands to facilitate chromosome identification has interfered and altered the banding pattern of *Rsa*I/Giemsa stain (Figure 3.5.6A to C). We have seen a number of uncertainties even after studying several individuals by this method. Chromosomes 1, 2, 7, 11, 12, 17, 19, and 20 show prominent bands at the centromeric regions (Table 3.5.2). The centromeric regions of chromosomes 9 and 16 show very small bands. The heterochromatic segment in the long arm of Y is not darkly stained under usual conditions. Some of the X chromosomes showed a minor dark band at the centromere. The short arms of acrocentric chromosomes show variable staining (Figure 3.5.7).

General Comments

The restriction enzymes are usually supplied as concentrated enzyme units in small quantities of respective buffer solution. These enzymes are diluted to the required working concentration (100 to 400 U per milliliter or as per requirement) with diluting solution (standard assay buffer) specific for each enzyme. Fresh slides not older than 1 week to 10 days are preferred for treatment with an enzyme. Even freshly prepared slides can be used. However, the finer details of the bands may be subdued and the chromosomes may acquire a ghostlike appearance even with slightly excessive treatment. The slides treated between the second and the fifth day after preparation yield the best differentiation. Preparations after the second day have a much higher resistance and tolerate a wider range in the duration of enzyme treatment, yet retain excellent morphologic characteristics.

The residual Giemsa bands are usually adequate for chromosome identification. However, if needed, the metaphases can be photographed using QFQ technique before treatment with the enzyme, but exposure to the light source should be kept to a minimum. Prolonged exposure may interfere or even block the effect of endonuclease, probably by forming the DNA-protein cross-linkage caused by radiation (Bianchi et al, 1984, 1985b). It should be mentioned that prior QFQ banding influences each enzyme treatment differently. For example, the distal long arm region of the Y chromosome and the pericentric region of the X chromosome tend to stain more darkly in prior Q-banded slides than in control slides.

Excessively treated slides, if not stained to the adequate intensity by Giemsa, can be observed and photographed using phase-contrast optics.

Mechanism

Chromosomal DNA in air-dried preparations is available for digestion with nucleases (Alfi et al, 1973; Sahasrabuddhe et al, 1978; Miller et al, 1983). Digestion with DNase I and micrococcal nuclease results in extensive cleavage of whole chromosomal DNA into small fragments of a few base pairs (bp) which are readily extracted. This results in an overall reduction in staining in metaphase. Conversely, endonucleases have short recognition sequences and cleave the DNA at specific sites, producing relatively larger fragments of DNA. The chromosomal regions that have a large number of recognition sites in their DNA for a particular endonuclease when digested with that enzyme would have smaller DNA fragments. Therefore, DNA is preferentially lost from these regions. The remaining regions, which have fewer recognition sites for the enzyme, lose negligible, if any, amounts of DNA. The loss of DNA is directly related to decreased Giemsa staining, which causes differential staining in metaphase chromosomes (Bianchi et al, 1985a; Mezzanotte et al, 1985; Miller et al, 1983). It is further suggested that DNA fragments longer than 1 kilobase pairs (kbp) would remain in the chromatin, whereas smaller fragments of about 200 bp long are extracted from the chro-

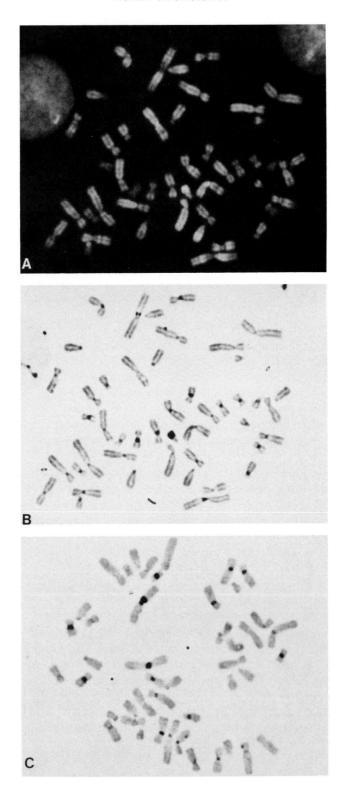

FIGURE 3.5.6. The effect of prior Q-banding on the staining pattern of *Rsa*I/Giemsa technique. A metaphase stained sequentially for Q-bands and for *Rsa*I/Giemsa-bands (A and B). Another cell stained by the *Rsa*I/Giemsa technique without prior Q-banding (C). The staining pattern in some of the chromosomes is altered by prior Q-banding.

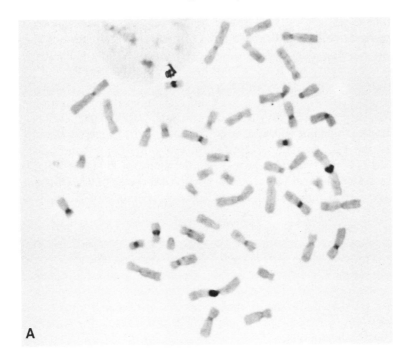

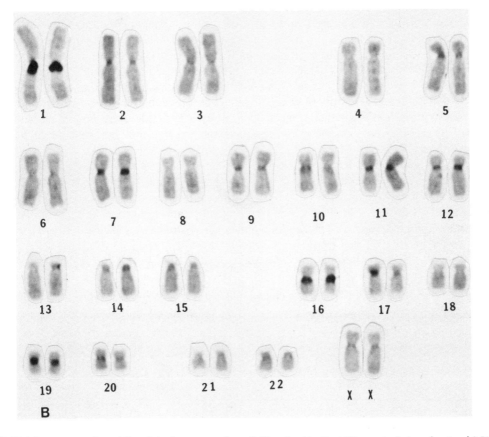

FIGURE 3.5.7. A metaphase (A) and the karyotype of a cell (B) stained by *Rsa*I/Giemsa technique (protocol 3.56).

mosomes (Miller et al, 1983, 1984). The endonuclease *Alu*I, which induces C-like bands, removes a significant amount (more than half) of the DNA from chromosomes except from certain C-band regions, whereas *Hae*III, which produces G- and C-like bands, removes a lesser amount (a quarter to less than half) of the DNA mostly from R-band regions of chromosomes (Mezzanotte et al, 1985; Miller et al, 1983).

Mezzanotte et al (1983b, 1985) suggested that the structural organization of chromatin, besides the frequency of restriction sites and extraction of DNA, may also play an important role in chromosome banding, because endonucleases *Eco*RI, *Hpa*II, *Hind*III, and *Bst*NI, which removed negligble amounts of DNA, have also shown G-like banding. It is worth mentioning however that the findings reported by Mezzanotte et al (1983b, 1985) and Miller et al (1983) with these endonucleases are inconsistent and therefore need further exploration.

It is adequately clear that chromatin condensation has no implications in endonuclease-induced bands (Bianchi et al, 1985a; Babu and Verma, 1986g).

Applications
The restriction endonuclease/Giemsa techniques are relatively much simpler to use and less time-consuming than are many others. These methods can be applied on fresh slides without aging and can be observed in a few hours. This feature is advantageous especially in prenatal diagnosis for rapid identification of unusual heteromorphisms (Rodriguez et al, 1986). The preparations are of a permanent nature and provide better resolution with very little distortion in chromosome morphology.

Some heteromorphisms, especially those of smaller magnitude, could not be delineated by C-banding. The heteromorphisms in chromosome 4 could be demonstrated by Q-banding and G-banding only when they are large enough to depict as a distinct band (McKenzie and Lubs, 1975; Bardhan et al, 1981; Docherty and Bowser-Riley, 1984). However, the *Alu*I/Giemsa technique is able to highlight even minor heteromorphisms that could easily be missed with other methods (Babu and Verma, 1986e). Chromosome 18 shows much smaller bands by *Alu*I/Giemsa technique than the corresponding C-bands, indicating that only a small portion of the C-band is resistant. This band is remarkably heteromorphic in size and position or even totally absent. This heteromorphic subset of C-band chromatin can be used as a chromosomal marker to identify the parental origin of additional chromosomes in the Edwards syndrome (Babu and Verma, 1986f). Although no specific study is available, other endonucleases such as *Dde*I and *Hin*fI may well be useful in the study of some heteromorphisms of human chromosomes.

C-like bands produced by endonuclease reveal a number of heterogeneous regions, possibly reflecting distinct entities of heterochromatin (Babu and Verma, 1986a). Instead of elaborate methods, enzymes can be applied to recognize areas of chromosomes containing repetitive DNA (Miller et al, 1983). The characteristic staining patterns are sometimes useful in identifying the origin of marker chromosomes (Babu et al, 1987b).

3.6 REFERENCES

Alfi OS, Donnel GN, Derencsenyi A: C-banding of human chromosomes produced by DNase. *Lancet* 1973;II:505.

Alhadeff B, Velivasakis M, Siniscalco M: Simultaneous identification of chromatid replication and of human chromosomes in metaphases of man-mouse somatic cell hybrids. *Cytogenet Cell Genet* 1977; 19:236–239.

Angelier N, Hernandez-Verdun D, Bouteille M: Visualization of Ag-NOR proteins on nucleolar transcriptional units in molecular spreads. *Chromosoma* 1982;86:661–672.

Arrighi FE, Hsu TE: Localization of heterochromatin in human chromosomes. *Cytogenetics* 1971;10: 81–86.

Babu KA, Verma RS: Fluorescent staining of heterochromatic regions of human chromosomes 1, 9, 15, 16, and the Y by DA/DAPI and D287/170 techniques. *Karyogram* 1984;10:28–30.

Babu A, Verma RS: Structural and functional aspects of nucleolar organizer regions (NORs) of human chromosomes. *Int Rev Cytol* 1985;94:151–176.

Babu A, Verma RS: Characterization of human chromosomal constitutive heterochromatin. *Can J Genet Cytol* 1986a;28:631–645.

Babu A, Verma RS: The heteromorphic marker on chromosome 18. *Am J Hum Genet* 1986b;38:549–554.

Babu A, Verma RS: Heteromorphic variants of chromosome 4. *Cytogenet Cell Genet* 1986c;41:60–61.

Babu A, Verma RS: Cytochemical heterogeneity of C-band in human chromosome 1. *Histochem J* 1986d;18:329–333.

Babu A, Macera MJ, Verma RS: Intensity heteromorphisms of human chromosome 15p by DA/DAPI technique. *Hum Genet* 1986e;73:298–300.

Babu A, Verma RS: Classification and incidence of pericentric heterochromatin of human chromosome 18. *Chromosoma* 1987a;95:163–166.

Babu A, Verma RS, Macera MJ: Identification of marker chromosomes by restriction endonucleases/Giemsa-technique in neoplastic cells. *Cancer Genet Cytogenet* 1987b;24:367–369.

Babu A, Verma RS: Chromosome structure, euchromatin and heterochromatin. *Int Rev Cytol* 1987c;108:1–60.

Bardhan S, Singh DN, Davis K: Polymorphism in chromosome 4. *Clin Genet* 1981;20:44–47.

Bianchi NO, Bianchi MS, Cleaver JE: The action of ultraviolet light on the patterns of banding induced by restriction endonucleases in human chromosomes. *Chromosoma* 1984;90:133–138.

Bianchi MS, Bianchi NO, Pantelias GE, Wolff S: The mechanism and pattern of banding induced by restriction endonucleases in human chromosomes. *Chromosoma* 1985a;91:131–136.

Bianchi NO, Morgan WF, Cleaver JE: Relationship of ultraviolet light-induced DNA-protein crosslinkage to chromatin structure. *Exp Cell Res* 1985b;156:405–418.

Bittman R: Studies of the binding of ethidium bromide to transfer ribonucleic acid: absorption, fluorescence, ultracentrifugation and kinetic investigations. *J Mol Biol* 1969;46:251–268.

Bloom SE, Goodpasture C: An improved technique for selective silver staining of nucleolar organizer regions in human chromosomes. *Hum Genet* 1976;34:199–206.

Bobrow M, Maden K, Pearson PL: Staining of some specific regions of human chromosomes, particularly the secondary constriction of No. 9. *Nature New Biol* 1972;238:122–124.

Bobrow M, Madan K: The effect of various banding procedures in human chromosomes studied with acridine orange. *Cytogenet Cell Genet* 1973a;12:145–156.

Bobrow M, Madan K: A comparison of chimpanzee and human chromosomes using the Giemsa-11 and other chromosome banding techniques. *Cytogenet Cell Genet* 1973b;12:107–116.

Bobrow M, Cross J: Differential staining of human and mouse chromosomes in interspecific cell hybrids. *Nature* 1974;251:77–79.

Brenner S, Pepper D, Berns MW, Tan E, Brinkley BR: Kinetochore structure, duplication, and distribution in mammalian cells: Analysis by human autoantibodies from scleroderma patients. *J Cell Biol* 1981;91:95–102.

Brinkley BR, Valdivia MM, Tousson A, Brenner SL: Compound kinetochores of the Indian muntjac. Evolution by linear fusion of unit kinetochores. *Chromosoma* 1984;91:1–11.

Brito-Babapulle V: Lateral asymmetry in human chromosomes 1, 3, 4, 15, and 16. *Cytogenet Cell Genet* 1981;29:198–202.

Buhler EM, Tsuchimoto T, Jurik L, Stalder G: Satellite DNA III and alkaline Giemsa staining. *Hum Genet* 1975;26:329–333.

Burgerhout W: Identification of interspecific translocation chromosomes in human-Chinese hamster hybrid cells. *Humangenetik* 1975;29:222–231.

Burkholder GD: The ultrastructure of R-banded chromosomes. *Chromosoma* 1981;83:473–480.

Burkholder GD: Silver staining of histone-depleted metaphase chromosomes. *Exp Cell Res* 1983;147:287–296.

Burkholder GD, Weaver MG: DNA-protein interactions and chromosome banding. *Exp Cell Res* 1977;110:251–262.

Busch H, Lischwe MA, Michalik J, Chan P-K, Busch RK: Nucleolar proteins of special interest: Silver-staining proteins B23 and C23 and antigens of human tomous nucleoli, in Jordan EG, Cullis CA (eds): *The Nucleolus*. New York, Cambridge University Press, 1982, pp 43–72.

Buys CHCM, Osinga J: Abundance of protein-bound sulfhydryl and disulfide groups at chromosomal nucleolus organizing regions: A cytochemical study on the selective silver staining of NORs. *Chromosoma* 1980;77:1–11.

Buys CHCM, Gouw WL, Blenkers JAM, van Dalen CH: Heterogeneity of human chromosome 9 constitutive heterochromatin as revealed by sequential distamycin A/DAPI staining and C-banding. *Hum Genet* 1981;57:28–30.

Caspersson T, Zech L, Johansson C, Modest EJ: Identification of human chromosomes by DNA-binding fluorescent agents. *Chromosoma* 1970;30:215–227.

Chandra P, Mildner B: Zur molekularen Wirkungsweise von Diaminphenylindols (DAPI). I. Physikochemische Untersuchungen zur Charakterisierung der Bindung von Diamidinphenylindol an die Nukleinsauren. *Cell Mol Biol* 1979;25:137–146.

Cherry LM, Johnston DA: Size variation in kinetochores of human chromosomes. *Hum Genet* 1987; 75:155–158.

Clark RJ, Felsenfeld G: Association of arginine-rich histones with GC-rich regions of DNA chromatin. *Nature (New Biol)* 1972;240:226–227.

Comings DE: Mechanisms of chromosome banding and implications for chromosome structure. *Ann Rev Genet* 1978;12:25–46.

Comings DE, Avelino E, Okada T, Wyandt HE: The mechanism of C- and G-banding of chromosomes. *Exp Cell Res* 1973;77:469–493.

Comings DE, Kavacs BW, Avelino E, Harris DC: Mechanisms of chromosome banding V. Quinacrine banding. *Chromosoma* 1975;50:115–145.

Comings DE, Drets ME: Mechanisms of chromosome banding. IX. Are variations in DNA base composition adequate to account for quinacrine, Hoechst 33258 and daunomycin banding? *Chromosoma* 1976;56:199–211.

Daniel A: Single Cd band in dicentric translocation with one suppressed centromere. *Hum Genet* 1979; 48:85–92.

Daniel A, Lam-Po-Tang PRLC: Mechanism for the chromosome banding phenomenon. *Nature* 1973; 244:358–359.

DeStefano GF, Ferrucci L: New cytogenetic techniques in the study of primate genome evolution. *Hum Genet* 1986;72:98–100.

Docherty Z, Bowser-Riley SM: A rare heterochromatic variant of chromosome 4. *J Med Genet* 1984; 21:470–472.

Donlon TA, Magenis RE: Structural organization of the heterochromatic region of human chromosome 9. *Chromosoma* 1981;84:353–363.

Donlon TA, Magenis RE: Methyl green is a substitute for distamycin A in the formation of distamycin A/DAPI C-bands. *Hum Genet* 1983;65:144–146.

Dutrillaux B, Lejeune J: Sur une novelle technique d'analyse du caryotype humain. *C R Acad Sci (D) (Paris)* 1971;272:2638–2640.

Earnshaw WC, Rothfield N: Identification of a family of human centromere proteins using autoimmune sera from patients with scleroderma. *Chromosoma* 1985;91:313–321.

Eiberg H: New selective Giemsa technique for human chromosomes, Cd staining. *Nature* 1974;248:55.

Ellison JR, Barr HJ: Quinacrine fluorescence of specific chromosome regions: Late replication and high AT content in *Samoaia leonensis*. *Chromosoma* 1972;36:375–390.

Erlanger BF, Beiser SM: Antibodies specific for ribonucleosides and ribonucleotides and their reaction with DNA. *Proc Natl Acad Sci USA* 1964;52:68–74.

Evans HJ, Ross A: Spotted centromeres in human chromosomes. *Nature* 1974;249:861–862.

Friend KK, Chen S, Ruddle FH: Differential staining of interspecific chromosomes in somatic cell hybrids by alkaline Giemsa stain. *Somatic Cell Genet* 1976;2:183–188.

Funaki K, Matsui S-I, Sasaki M: Location of nucleolar organizer in animal and plant chromosomes by means of an improved N-banding technique. *Chromosoma* 1975;49:357–370.

Gagne R, Laberge C: Specific cytological recognition of the heterochromatic segment of number 9 chromosome in man. *Exp Cell Res* 1972;73:239–242.

Gill JE, Jotz MM, Young SG, Modest EJ, Sengupta SK: 7-amino-actinomycin D as a cytochemical probe. *J Histochem Cytochem* 1975;23:793–799.

Goodpasture C, Bloom SE: Visualization of nucleolar organizer regions III. Mammalian chromosomes using silver staining. *Chromosoma* 1975;53:37–50.

Goodpasture C, Bloom SE, Hsu TC, Arrighi FE: Human nucleolus organizers: The satellites or the stalks? *Am J Hum Genet* 1976;28:559–566.

Gosden JR, Spowart G, Lawrie SS: Satellite DNA and cytological staining patterns in heterochromatic inversions of human chromosome 9. *Hum Genet* 1981;58:276–278.

Hays T, Morse HG, Robinson A: 9;22;15 complex translocation in Ph[1] chromosome positive CML revealed by Giemsa 11 procedure in apparent lymphoid cells of blastic crisis. *Cancer Genet Cytogenet* 1981;4:283–292.

Howell WM, Black DA: Controlled silver-staining of nucleolus organizer regions with a protective colloidal developer: A 1-step method. *Experientia* 1980;36:1014.

Howell WM, Hsu TC: Chromosome core structure revealed by silver staining. *Chromosoma* 1979;73: 61–66.

Hsieh T, Brutlag DL: A protein that preferentially binds Drosophila satellite DNA. *Proc Natl Acad Sci USA* 1979;76:726-730.

Hsu TC, Pathak S, Chen TR: The possibility of latent centromeres and a proposed nomenclature system for total chromosome and whole arm translocations. *Cytogenet Cell Genet* 1975;15:41-49.

Hubbell HR, Hsu TC: Identification of nucleolus organizer regions (NORs) in normal and neoplastic human cells by the silver-staining technique. *Cytogenet Cell Genet* 1977;19:185-196.

ISCN: *An International System for Human Cytogenetics Nomenclature: Birth Defects.* Original Article Series, 1985 Vol 21, No.1 March of Dimes Birth Defects Foundation, New York.

Jensen RH, Langlois RG, Mayall BH: Strategies for choosing a deoxyribonucleic acid stain for flow cytometry of metaphase chromosomes. *J Histochem Cytochem* 1977;25:954-964.

Jones KW: Repetitive DNA and primate evolution, in Yunis, JJ (ed): *Molecular Structure of Human Chromosomes.* New York, Academic Press, 1977, pp 295-326.

Jorgenson KF, van de Sande JH, Lin CC: The use of base pair specific DNA binding agents as affinity labels for the study of mammalian chromosomes. *Chromosoma* 1978;68:287-302.

Kaelbling M, Miller DA, Miller OJ: Restriction enzyme banding of mouse metaphase chromosomes. *Chromosoma* 1984;90:128-132.

Latt SA, Wohlleb J: Optical studies of the interaction of 33258 Hoechst with DNA, chromatin and metaphase chromosomes. *Chromosoma* 1975;52:297-316.

Latt SA, Brodie S, Munroe SJ: Optical studies of complexes of quinacrine with DNA and chromatin: Implications for the fluorescence of cytological chromosome preparations. *Chromosoma* 1974;49:17-40.

Latt SA, Sahar E, Eisenhard ME: Pairs of fluorescent dyes as probes of DNA and chromosomes. *J Histochem Cytochem* 1979;27:65-71.

Latt, SA, Juergens LA, Matthews DJ, Gustashaw KM, Sahar E: Energy transfer-enhanced chromosome banding. An overview. *Cancer Genet Cytogenet* 1980;1:187-196.

Leemann U, Ruch F: Selective excitation of mithramycin or DAPI fluorescence on double-stained cell nuclei and chromosomes. *Histochemistry* 1978;58:329-334.

Lima-de-Faria A, Isaksson M, Olsson E: Action of restriction endonucleases on the DNA and chromosomes of *Muntiacus muntjak. Hereditas* 1980;92:267-273.

Lin MS, Comings DE, Alfi OS: Optical studies of the interaction of 4′-6-diamidino-2-phenylindole with DNA and metaphase chromosomes. *Chromosoma* 1977;60:15-25.

Lober G, Beensen V, Zimmer Ch, Hanschmann H: Changes of quinacrine staining on human chromosomes by the competitive binding of A-T and G-C specific substances. *Studia Biophysica* 1978;69:237-238.

Magenis RE, Donlon TA, Wyandt HE: Giemsa-11 staining of chromosome 1: A newly described heteromorphism. *Science* 1978;202:64-65.

Maraschio P, Zuffardi O, Curto FL: Cd bands and centromeric function in dicentric chromosomes. *Hum Genet* 1980;54:265-267.

Matsui S-I, Sasaki M: Differential staining of nucleolus organisers in mammalian chromosomes. *Nature* 1973;246:148-150.

Mattie M-G, Mattei J-F, Ayme S, Giraud F: Dicentric Robertsonian translocations in man (17 cases studied by R, C and N banding). *Hum Genet* 1979;50:33-38.

Mattei M-G, Mattei J-F, Vidal I, Giraud F: Advantages of silver staining in seven rearrangements of acrocentric chromosomes, excluding Robertsonian translocations. *Hum Genet* 1980;54:365-370.

McKay RDG: The mechanism of G and C banding in mammalian metaphase chromosome. *Chromosoma* 1973;44:1-14.

McKenzie WH, Lubs HA: Human Q and C chromosomal variants: Distribution and incidence. *Cytogenet Cell Genet* 1975;14:97-115.

Merry DE, Pathak S, Hsu TC, Brinkley BR: Anti-kinetochore antibodies: Use as probes for inactive centromeres. *Am J Hum Genet* 1985;37:425-430.

Mezzanotte R, Ferrucci L, Vanni R, Bianchi U: Selective digestion of human metaphase chromosomes by *Alu*I restriction endonuclease. *J Histochem Cytochem* 1983a;31:553-556.

Mezzanotte R, Bianchi U, Vanni R, Ferrucci L: Chromatin organization and restriction nuclease activity on human metaphase chromosomes. *Cytogenet Cell Genet* 1983b;36:562-566.

Mezzanotte R, Ferrucci L, Vanni R, Summner AT: Some factors affecting the action of restriction endonucleases on human metaphase chromosomes. *Exp Cell Res* 1985;161:247-253.

Mikelsaar A-V, Schwarzacher GG, Schnedl W, Wagenbichler P: Inheritance of Ag-stainability of nucleolus organizer regions: Investigation in 7 families with trisomy 21. *Hum Genet* 1977;38:183-188.

Mikkelsen M, Basli A, Poulsen H: Nucleolus organizer regions in translocations involving acrocentric chromosomes. *Cytogenet Cell Genet* 1980;26:14-21.

Miller DA, Dev VG, Tantravahi R, Miller OJ: Suppression of human nucleolus organizer activity in mouse-human somatic hybrid cells. *Exp Cell Res* 1976;101:235–243.

Miller DA, Choi Y-C, Miller OJ: Chromosome localization of highly repetitive human DNA's and amplified ribosomal DNA with restriction enzymes. *Science* 1983;219:395–397.

Miller DA, Gosden JR, Hastie ND, Evans HJ: Mechanism of endonuclease banding of chromosomes. *Exp Cell Res* 1984;155:294–298.

Miller OJ, Miller DA, Warburton D: Application of new staining techniques to the study of human chromosomes, in Steinberg AG, Bearn A (eds): *Progress in Medical Genetics*. Philadelphia, WB Saunders, 1973, vol 9, pp 1–47.

Miller OJ, Schnedl W, Allen J, Erlanger BF: 5-Methylcytosine localized in mammalian constitutive heterochromatin. *Nature* 1974;251:636–637.

Miller OJ, Miller DA, Dev VG, Tantravahi R, Croce CM: Expression of human and suppression of mouse nucleolus activity in mouse-human somatic cell hybrids. *Proc Natl Acad Sci USA* 1976;73: 4531–4535.

Miller OJ, Miller DA, Tantravahi R, Dev VG: Nucleolus organizer activity and the origin of Robertsonian translocations. *Cytogenet Cell Genet* 1978;20:40–50.

Moroi Y, Peebles C, Fritzler MJ, Steigerwald J, Tan EM: Autoantibody to centromere (kinetochore) in scleroderma sera. *Proc Natl Acad Sci USA* 1980;77:1627–1631.

Morse HG, Hays T, Patterson D, Robinson A: Giemsa-11 technique. Applications in the chromosomal characterization of hematologic specimens. *Hum Genet* 1982;61:141–144.

Muller W, Crothers DM: Studies of the binding of actinomycin and related compounds to DNA. *J Mol Biol* 1968;35:251–290.

Muller W, Gautier F: Interactions of heteroaromatic compounds with nucleic acids. A-T-specific non-intercalating DNA ligands. *Eur J Biochem* 1975;54:385–394.

Muller W, Bunemann H, Dattagupta N: Interaction of heteroaromatic compounds with nucleic acids. 2. Influence of substituents on the base and sequence specificity of intercalating ligands. *Eur J Biochem* 1975;54:279–291.

Nakagome Y, Teramura F, Kataoka K, Hosono F: Mental retardation, malformation and partial 7 p monosomy 45,XX,tdic(7;15)(p21;p11). *Clin Genet* 1976;9:621–634.

Nakagome Y, Abe T, Misawa S, Takeshita T, Iinuma K: The "loss" of centromeres from chromosomes of aged women. *Am J Hum Genet* 1984;36:398–404.

Okamoto E, Miller DA, Erlanger BF, Miller OJ: Polymorphism of 5-methylcytosine-rich DNA in human acrocentric chromosomes. *Hum Genet* 1981;58:255–259.

Olson SB, Magenis RE, Lovrien EW: Human chromosome variation: The discriminatory power of Q-band heteromorphism (variant) analysis in distinguishing between individuals, with specific application to cases of questionable paternity. *Am J Hum Genet* 1986;38:235–252.

Paris Conference: Standardization in human cytogenetics. Birth Defects: Original Article Series, 1971, Vol 8, No. 7. New York, The National Foundation, 1972; also in *Cytogenetics* 1971;11:313–362.

Paris Conference, Supplement: Standardization in human cytogenetics: Birth Defects: Original Article Series, 1975; Vol 11, No. 9, The National Foundation, New York, 1975; also in *Cytogenet Cell Genet* 1975;15:201–238.

Pachmann U, Rigler R: Quantum yield of acridines interacting with DNA of defined base sequence. *Exp Cell Res* 1972;72:602–608.

Pathak S, Arrighi FE: Loss of DNA following C-banding procedures. *Cytogenet Cell Genet* 1973;12: 414–422.

Pathak S, Hsu TC: Silver-stained structures in mammalian meiotic prophase. *Chromosoma* 1979;70: 195–203.

Rigler R: Microfluorometric characterization of intracellular nucleic acids and nucleoproteins by acridine orange. *Acta Physiol Scand* 1966;67(Suppl 267):1–122.

Rodriguez JG, Babu A, Verma RS: A rapid method of culturing bloody amniotic fluid for chromosome analysis. *Am J Obstet Gynecol* 1986;154:969–970.

Romain DR, Columbano-Green L, Sullivan J, Smythe RH, Gebbie O, Rarfitt R, Chapman C: Cd-banding studies in a homologous Robertsonian 13;13 translocation. *J Med Genet* 1982;19:306–310.

Roos U-P: Are centromeric dots kinetochores? *Nature* 1975;254:463.

Sahar E, Latt SA: Enhancement of banding patterns in human metaphase chromosomes by energy transfer. *Proc Natl Acad Sci USA* 1978;75:5650–5654.

Sahar E, Latt SA: Energy transfer and binding competition between dyes used to enhance staining differentiation in metaphase chromosomes. *Chromosoma* 1980;79:1–28.

Sahasrabuddhe CG, Pathak S, Hsu TC: Responses of mammalian metaphase chromosomes to endonuclease digestion. *Chromosoma* 1978;69:331–338.

Schmid M, Muller HJ, Stasch S, Engel W: Silver staining of nucleolus organizer regions during human spermatogenesis. *Hum Genet* 1983;64:363–370.

Schnedl W, Dev VG, Tantravahi R, Miller DA, Erlanger BF, Miller OJ: 5-Methylcytosine in heterochromatic regions of chromosomes: Chimpanzee and gorilla compared to the human. *Chromosoma* 1975; 52:59–66.

Schnedl W, Dann O, Schweizer D: Effects of counterstaining with DNA binding drugs on fluorescent banding patterns of human and mammalian chromosomes. *Eur J Cell Biol* 1980;20:290–296.

Schnedl W, Abraham R, Dann O, Geber G, Schwizer D: Preferential fluorescent staining of heterochromatic regions in human chromosomes 9, 15, and the Y by D287/170. *Hum Genet* 1981;59:10–13.

Schreck RR, Erlanger BF, Miller OJ: Binding of anti-nucleoside antibodies reveals different classes of DNA in the chromosomes of the kangaroo rat (*Dipodomys ordii*). *Exp Cell Res* 1977;108:403–411.

Schreck RR, Erlanger BF, Miller OJ: The use of anti-nucleoside antibodies to probe the organization of chromosomes denatured by ultraviolet irradiation. *Exp Cell Res* 1974;88:31–39.

Schreck RR, Breg WR, Erlanger BF, Miller OJ: Preferential derivation of abnormal human G-group-like chromosomes from chromosome 15. *Hum Genet* 1977;36:1–12.

Schuh BE, Korf BR, Salwen MJ: Dynamic aspects of trypsin-Giemsa banding. *Humangenetik* 1975; 28:233–238.

Schwarzacher HG, Mikelsaar A-V, Schnedl W: The nature of the Ag-staining of nucleolus organizer regions. Electron- and light-microscopic studies on human cells in interphase, mitosis, and meiosis. *Cytogenet Cell Genet* 1978;20:24–39.

Schweizer D: Reverse fluorescent chromosome banding with chromomycin and DAPI. *Chromosoma* 1976;58:307–324.

Schweizer D: R-banding produced by DNase I digestion of chromomycin-stained chromosomes. *Chromosoma* 1977;64:117–124.

Schweizer D: Simultaneous fluorescent staining of R bands and specific heterochromatic regions (DA/DAPI bands) in human chromosomes. *Cytogenet Cell Genet* 1980;27:1980–193.

Schweizer D: Counterstain-enhanced chromosome banding. *Hum Genet* 1981;57:1–14.

Schweizer D, Ambros P, Andrle M: Modification of DAPI banding on human chromosomes by prestaining with a DNA-binding oligopeptide antibiotic, distamycin A. *Exp Cell Res* 1978;111:327–332.

Sehested J: A simple method for R-banding of human chromosome showing a pH-dependent connection between R and G bands. *Humangenetik* 1974;21:55–58.

Stenman S, Rosenqvist M, Ringertz NR: Preparation and spread of unfixed metaphase chromosomes for immunofluorescence staining of nuclear antigens. *Exp Cell Res* 1975;90:87–94.

Sumner AT: A simple technique for demonstrating centromeric heterochromatin. *Exp Cell Res* 1972; 75:304–306.

Sumner AT: The nature and mechanisms of chromosome banding. *Cancer Genet Cytogenet* 1982;6: 59–87.

Szafran H, Beiser SM, Erlanger BF: The use of egg albumin conjugates for the purification of antibody. *J Immunol* 1969;103:1157–1158.

van de Sande JH, Lin CC, Jorgenson KF: Reverse banding on chromosomes produced by a guanosine-cytosine specific DNA binding antibiotic: Olivomycin. *Science* 1977;195:400–402.

VedBrat S, Verma RS, Dosik H: NSG banding of sequentially QFQ and RFA banded human acrocentric chromosomes. *Stain Technol* 1980;55:77–80.

Verma RS: The varieties of R-banding: Their methodology and application. *Karyogram* 1984;8:72–73.

Verma RS, Dosik H: Human chromosomal heteromorphisms: Nature and clinical significance. *Int Rev Cytol* 1980;62:361–383.

Verma RS, Lubs HA: A simple R banding technique. *Am J Hum Genet* 1976;27:110–117.

Verma RS, Lubs HA: Additional observations on the preparations of R-banded human chromosomes with acridine orange. *Canadian J Genet Cytol* 1976;18:45–50.

Verma RS, Dosik H, Lubs HA: Size and pericentric inversion heteromorphisms of secondary constriction regions (h) of chromosomes 1, 9 and 16 as detected by CBG technique in caucasians. *Am J Med Genet* 1979;2:331–339.

Verma RS, VedBrat S, Dosik H: A rapid method for identification of constitutive heterochromatin of secondary constriction regions of human chromosomes 1, 9, and 16. *J Heredity* 1982;73:74–76.

Verma RS, Babu KA, Rosenfeld W, Jhaveri RC: Marker chromosome in "cat eye syndrome." *Clin Genet* 1985;27:526–528.

Weisblum B, deHaseth PL: Nucleotide specificity of the quinacrine staining reaction for chromosomes. *Chromosomes Today* 1973;4:35–51.

Williams MA, Kleinschmidt JA, Krohne G, Franke WW: Argyrophilic nuclear and nucleolar proteins of Xenopus laevis oocytes identified by gel electrophoresis. *Exp Cell Res* 1982;137:341–351.

Wisniewski LP, Doherty RA: Supernumerary microchromosomes identified as inverted duplications of chromosome 15: A report of three cases. *Hum Genet* 1985;69:161–163.

Zimmer C: Effects of the antibiotic netropsin and distamycin A on the structure and function of nucleic acids. *Prog Nucleic Acid Res Mol Biol* 1975;15:285–318.

Specialized Techniques

4.1 HIGH RESOLUTION BANDING

Since the inception of banding methods to facilitate the unequivocal identification of chromosomes, major interest has been devoted to obtaining less condensed chromosomes. An advantage of elongated chromosomes is that they can visualize finer details with a large number of bands that otherwise are represented as a few major bands in condensed metaphase chromosomes. A simple way to achieve this is to curtail the exposure of cells to colcemid or other compounds used to arrest the cells at metaphase or to reduce their concentration. These variations have certain limitations in practice. Mitotic index is significantly low if exposure time is reduced; and chromosome spreading is hindered if the cultures are exposed to too low concentrations of colcemid. Therefore, an alternative approach to accumulate a maximum number of mitoses by synchronizing the cells and exposing them briefly to appropriate concentrations of colcemid is adopted to obtain considerably long chromosomes.

Protocol 4.11.1 Methotrexate (MTX)/Thymidine Method

(Following Yunis, 1976)

Solutions

1. Methotrexate (MTX) solution

   MTX stock solution (1×10^{-3} M)

MTX [(+)Amethopterin]	1 mg
Distilled water	2.2 ml

   MTX working solution (1×10^{-5} M)

MTX stock solution (1×10^{-3} M)	0.1 ml
Distilled water	9.9 ml

   MTX is relatively unstable and can be used for a week or two. Store frozen in dark vials. Sterilization is usually not required.

2. Thymidine solution (1×10^{-3} M)

Thymidine	2.5 mg
Distilled water	10 ml

   Filter sterilize and store frozen in small aliquots. Solution is relatively stable and can be used for 2 to 3 months.

Procedure

1. Set up the peripheral blood cultures following either the macroculture or microculture method described in protocols 2.11.2 or 2.12.1, respectively.

2. Incubate the cultures at 37°C for 72 hours.

3. Add 0.1 ml of MTX working solution (1×10^{-5} M) to each culture consisting of 10 ml of medium. (Final concentration of MTX in culture medium is 1×10^{-7} M.)

4. Incubate the cultures in the dark at 37°C for 17 hours (overnight).

5. Centrifuge the cultures at 800 rpm for 8 minutes and discard the medium containing MTX.

6. Wash the cells twice by suspending them in fresh unsupplemented medium (medium without serum), by centrifuging (800 rpm for 8 minutes) and then discarding the supernatant each time. After the final wash, suspend the cells in complete culture medium containing serum and antibiotics at usual concentrations.

7. Add 0.1 ml of thymidine solution (1×10^{-3} M) to each culture containing 10 ml of medium. (Final concentration of thymidine in culture is 1×10^{-5} M.)

8. Incubate the cultures at 37°C for 5.5 to 6 hours.

9. Add 0.02 ml (20 μl) of colcemid (10 μg/ml) to 10 ml of culture medium and incubate for an additional 10 to 15 minutes.

10. Harvest the cultures following routine protocol 2.11.3.

11. Chromosome preparations are banded according to protocol 3.12.1 for G-bands.

Comments

Overnight treatment of cultures with MTX is followed, that is, add MTX to cultures before leaving work and wash the cells the following morning. Incubation time before MTX treatment can be cut down to 48 hours instead of 72 hours (step 2) to accommodate the work schedule. Cultures should be incubated in the dark after MTX has been added. Colcemid treatment should be kept to a minimum to obtain chromosomes that are least condensed (longer chromosomes). Incubation time after releasing the mitotic block should be at least 5 hours or more and can be extended to 10 hours. The hypotonic treatment may be slightly increased to facilitate thorough spreading of chromosomes and to increase the number of usable metaphases.

Protocol 4.11.2   MTX/Bromodeoxyuridine (BrdU) Method

(Following Pai and Thomas, 1980)

Solutions

1. MTX working solution (1×10^{-5} M)

   Same as in protocol 4.11.1

2. Fluorodeoxyuridine (FUdR) stock solution (4×10^{-4} M)

FUdR	1 mg
Distilled water	10 ml

   Store frozen in dark vials. The solution can be used for 3 to 4 weeks.

3. Uridine stock solution (3×10^{-3} M)

Uridine	1 mg
Distilled water	1 ml

   Store frozen in small vials.

4. Bromodeoxyuridine (BrdU) solution

BrdU	3 mg	(10^{-2} M)
FUdR stock solution (4×10^{-4} M)	0.1 ml	(4×10^{-5} M)
Uridine (6×10^{-3})	0.2 ml	(6×10^{-4} M)
Distilled water	0.9 ml	

   Store frozen in dark vials. Filter sterilization is generally not required. The solution can be used over a period of 2 weeks.

Procedure

1. Set up the lymphocyte cultures following the routine protocols and incubate for 48 or 72 hours at 37°C.

2. Add 0.1 ml of MTX working solution to each culture containing 10 ml of medium. Incubate the cultures for 17 hours or overnight.

3. Wash the cells twice in serum-free medium to remove MXT from the cultures as described in step 6 of previous protocol 4.11.1.

4. After suspending the cells in regular culture medium, add 0.1 ml of BrdU solution to each 10-ml culture medium.

5. Incubate the cultures for an additional 5.5 to 6 hours.

6. Add 0.02 ml (20 μl) of colcemid and incubate for another 10 to 15 minutes.

7. Harvest follows the routine protocol.

8. Stain the slides for R-bands according to protocol 3.15.1 or 3.16.1.

Comments
See comments under the previous protocol 4.11.1.

General Comments
The synchronization protocols to obtain high resolution bands can virtually be adopted for any type of cell culture. These methods can be used for short-term bone marrow cultures without procedural changes. However, it is important to know that the blast cells in many bone marrow samples have a longer cell cycle and therefore require a relatively longer incubation time (7 to 11 hours) after the mitotic block is released (Gallo et al, 1984; Webber and Garson, 1983). Monolayer cultures can be synchronized principally the same way by adding an appropriate amount of MTX working solution for a period of 17 hours or overnight. The mitotic block is released by rinsing the monolayers in two or three changes of medium and replacing with culture medium. Thymidine or BrdU should be added to the medium as desired (Cheung et al, 1985). It should also be noted that if tumor cell cultures are synchronized, they would require prolonged colcemid treatment to obtain adequate preparations.

Other chemical agents like FUdR can also be used at concentrations of 0.1 μM (supplemented with uridine) successfully to block the mitotic cycle and synchronize the cells (Webber and Garson, 1983).

Applications
Cultures harvested by synchronization methods yield higher mitotic indices than do those harvested by routine protocols. This in turn allows the length of colcemid treatment to be shortened to obtain less condensed chromosomes of prophase and prometaphase stages. Many bands can be seen in chromosomes from such preparations (Figure 4.1.1).

High resolution banding is now used routinely in several laboratories for thorough investigation of clinical cases and gene localization (Yunis et al, 1978). It has been documented that chromosomal abnormalities in leukemia can be better resolved using these high resolution techniques (Yunis, 1981a, b; Yunis et al, 1981; 1984).

4.2   CHROMOSOMAL FRAGILE SITES

Although the marker X chromosome was described by Lubs in 1968, it was not until a decade ago that expression of the fragile site in the X chromosome in response to culture conditions received major attention. The fragile sites are nonstaining gaps usually involving both chro-

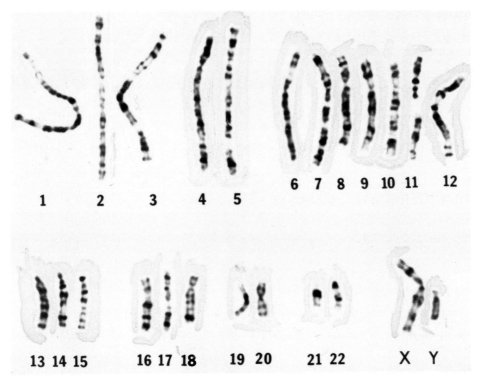

FIGURE 4.1.1. Selected haploid set of chromosomes obtained by high resolution technique (stained by GTG technique).

matids present at the same point in the chromosome in a significant number of cells and inherited in a dominant Mendelian fashion. These are generally referred to as "heritable" or "rare" fragile sites and should be distinguished from nonheritable "common" fragile sites which may appear as a result of various treatments (Sutherland, 1979). Although the majority of fragile sites are expressed in response to external culture conditions, at least one in the long arm of chromosome 16(q22) in some individuals seems to be independent of culture conditions. Among the known fragile sites, the one in the long arm of the X chromosome (Xq27 or Xq28, see general comments) is of the utmost importance because of its well-established relation to a particular type of mental retardation, the Martin Bell syndrome. So far, the other fragile sites apparently have no known clinical significance (described later in this section). The peripheral blood lymphocytes of suspected patients and carriers are now routinely investigated for fragile X [fra(X)] chromosome in a number of clinical laboratories.

4.21   FRAGILE X [FRA(X)(q27.3)]

The rare fragile site in X chromosome (q 27.3) is rarely expressed in normal culture conditions, but it is dependent on a number of external factors. Therefore, blood samples of the patients in question have to be cultured by methods that promote expression of the fragile site in the X chromosome. They include either cultivating the cells in folate-deficient medium or treating the cells with thymidylate synthetase inhibitors such as fluorodeoxyuridine (FUdR) or methotrexate (MTX) in culture. Commercially available media usually have folic acid. One can obtain folic-acid-free medium especially formulated for this purpose but it is not always essential. Any one of the following methods or other variations can be used to study fra(X).

Protocol 4.21.1   Using Folate-Deficient Medium

(Following Sutherland, 1979)

Solutions

1. Culture medium

Medium 199 (HEPES buffered)	100 ml
Fetal bovine serum	5 ml
Penicillin + streptomycin	1 ml
L-glutamine	1 ml

Medium 199 is used because of its low folic acid content. Prepared medium is stored at 2° to 4°C. Medium is good to use up to 2 to 3 weeks.

Procedure

1. Set up the peripheral blood cultures using the aforementioned medium following either macroculture (separated lymphocytes) (protocols 2.21.1 and 2.21.2) or microculture (whole blood) technique (protocol 2.22.1).
2. Incubate the culture tubes in a slanting position at 37°C for 92 hours (4 days).
3. Add colcemid to each culture tube at a final concentration of 0.1 μg/ml 1 hour before harvest.
4. Harvest the cultures following routine protocol 2.21.3.
5. Stain the slides for unbanded or G-banded chromosomes as desired (see general comments).

Technical Comments

The culture medium can be prepared using medium 199 with Earle's buffer instead of 199 with HEPES buffer. However, HEPES buffered medium is preferred over that of Earle, because no further care of cultures, regarding pH, is required until harvest time. If Earle's buffered medium is used, cultures should be adjusted to and maintained in alkaline (pH 7.4 to 7.6) condition especially during the final hours before harvesting. The pH of the culture medium can be made alkaline by loosening the caps of the culture vessel, to release CO_2, for a few hours (this will not help if cultures are kept in a CO_2 incubator). The concentration of serum may further be reduced to 2% instead of 5%.

Protocol 4.21.2   Using Thymidylate Synthetase Inhibitor

(Following Glover, 1981; Mattei et al, 1981)

Solutions

1. Culture medium
   Same as in protocol 2.22.1.
2. Thymidylate synthetase inhibitor solution
   FUdR solution
   FUdR stock solution (1 mM)

FUdR	1 mg
Distilled water	4 ml

FUdR working solution (10 μM)

FUdR stock solution	0.1 ml
Distilled water	9.9 ml

Store solution frozen in dark vials. Filter sterilization is usually not essential. Prepare fresh FUdR solution each week.

OR

MTX solution (1 mg/ml)

MTX [(+) amethopterin]	10 mg
Distilled water	10 ml

Store solution frozen in dark vials. No filter sterilization is usually required. Prepare fresh solution each week.

Procedure

1. Set up the cultures of peripheral blood using either macroculture or microculture technique in a regular culture medium.

2. Incubate the cultures at 37°C for 72 hours.

3. Add *either* 0.1 ml of FUdR working solution (10 μM) to each culture containing 10 ml of medium (final concentration 0.1 μM) *or* 0.1 ml of MTX solution (1 mg/ml) to 10 ml culture medium (final concentration 10 μg/ml).

4. Incubate the cultures in the dark for an additional 24 hours at 37°C. (Tubes should either be wrapped in a black wrapper or kept in total darkness.)

5. Harvest the cultures following routine protocol 3.12.1 with colcemid for a final 1 hour.

6. Slides may be stained for unbanded or banded chromosomes.

Technical Comments

In this protocol, the concentration of serum is not a critical factor for the expression of fra(X) and could be up to 20%. Both FUdR and MTX are toxic to the cells, and higher concentrations would reduce the mitotic index significantly. pH is also an important factor and should be alkaline (usually between pH 7.2 and 7.6) preferably throughout the culture period and is particularly important during the final hours before harvesting. The expression of fra(X) is influenced by cell density. Cultures with high cell density express fra(X) in a smaller number of cells and vice versa (Krawczun et al, 1986). Therefore, lymphocyte cultures should be initiated with a relatively low cell number. Some studies have used bromodeoxyuridine (BrdU) incorporation to obtain direct R-bands by RBA or RBG procedures. However, the addition of BrdU to cultures even in the presence of either thymidylate synthetase inhibitor drastically reduces the expression of fra(X).

General Comments

Confusion exists with respect to the band involved in fra(X) chromosome. It has often been described as either q27 or q28 apparently because of the difference in interpretation rather than the heterogeneity in the position of fragile site among different patients. In more recent reports, it is referred to as q27.3.

The number of cells exhibiting fragile site in the long arm of X chromosome varies significantly in different subjects who are positive for fra(X). The variation may depend to a certain extent on the technique used. Fragile sites manifest in a variety of ways. Some of the expressions using different staining procedures are shown in Figures 4.2.1A and B. Unbanded chromosomes apparently are the most suitable for analysis of fragile sites in metaphase chro-

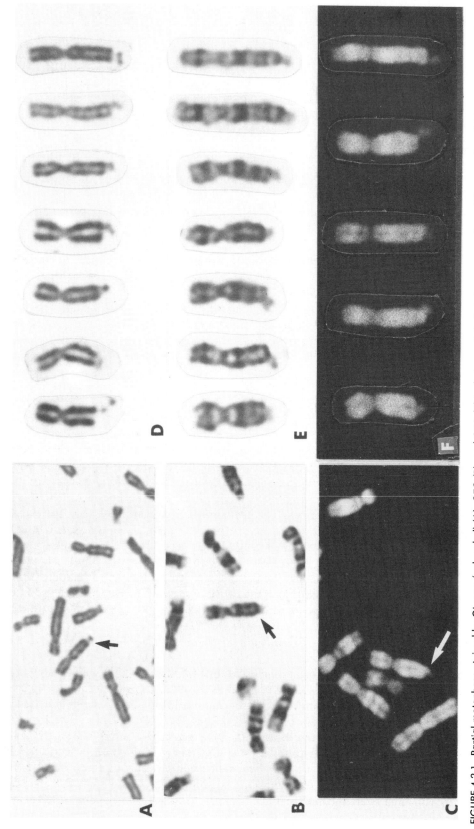

FIGURE 4.2.1. Partial metaphase stained by Giemsa (unbanded) (A), GTG (B), and QFQ (C) techniques. Selected chromosomes from different cells exhibiting various manifestations of fra(x) with respective staining techniques (D, E, and F).

mosomes. Unbanded preparations, however, may create confusion, at least in some cells, in ascertaining that the chromosome bearing a fragile site is in fact the X chromosome and not any other C-group autosome. This problem can be resolved by subsequently examining the same preparations stained by either the QFQ or the GTG technique. Alternatively, analysis of fra(X) can be performed directly using banded chromosomes (Figure 4.2.1, middle and bottom rows). Chromosomes stained by fluorochromes are the least suitable, and several cells with fra(X) could be missed during microscopic examination (Figure 4.2.1, bottom row).

In X chromosome, besides the fra(X) (q27.3), two common fragile sites are expressed by aphidicolin treatment. In addition, a coincident common fragile site, in response to aphidicolin treatment, has recently been described at the same region of the X chromosome (Table 4.2.1).

Analysis of fra(X)

It is essential for any laboratory to standardize and periodically test one or preferably two protocols before evaluation of clinical patients. The protocols should be tested by processing a positive control sample from a previously known patient or using a lymphoblastoid cell line established from a fra(X) patient. Without an established proven protocol, the negative findings in samples of suspected patients would be questionable. It is a good practice to follow more than one protocol to process each sample, one with medium consisting of low serum and the other involving thymidylate synthetase inhibitor (for final 24 hours) treatment. A total of 100 cells should generally be examined for the fra(X) chromosome.

If the preparations exhibit fra(X) in a significant number of cells, the patient is considered to be positive. However, ascertainment becomes critical when analysis reveals a low percentage, for example, 1 or 2%, of cells showing fra(X). Although a 4% cutoff point was suggested, no definite system has yet been adopted to discriminate between positive and negative cases (Jacobs et al, 1980). In these cases, it is necessary to supplement the study by repeating the test in that particular patient and/or by investigating other affected or possible carrier members of the family to ascertain conclusively the status of fra(X).

The number of cells showing fra(X) in a given patient seems to be influenced by the technique used (Lubs et al, 1984). The frequency of cells exhibiting fra(X) varies to a great extent in different individuals. The rate of fra(X) is consistent over time and in replicate cultures of an individual (Jenkins et al, 1986). Soudek et al (1984) noted that the frequency of fra(X) is similar in related members of a family under the same experimental conditions. These findings suggest that the rate of expression of a fragile site possibly is characteristic of a particular X chromosome and familially inherited.

Prenatal Diagnosis of fra(X)

A limited number of studies on prenatal diagnosis of fra(X) are available (Jenkins et al, 1981, 1984; Sutherland et al, 1987; Tommerup et al, 1981, Webb et al, 1987). The fetal tissues used in these studies include amniotic fluid, fetal blood obtained by fetoscopy, and skin obtained after the termination of pregnancy. The cells are grown in either medium 199, McCoy's 5A, Ham's F10, or RPMI 1640 supplemented with 10% amniotic fluid or 20% fetal calf serum and exposed to FUdR, MTX, or excess thymidine for 24 hours before harvest. These studies are performed exclusively in conceptions of patients who are known to be carriers or who have a strong family history suggesting a carrier status. One should be aware of the difficulties and observe extreme caution in prenatal diagnosis of fra(X). At the present time it is still in an experimental stage and is not offered as a routine clinical test.

Mechanism

Since the early descriptions of fragile sites, the expression of fragile site on X chromosome is closely linked to the folate metabolism. The presence of folate is an essential requirement for growth of mammalian cells in culture. Fra(X) is expressed in response to deficiency of

Table 4.2.1. Fragile Sites in Human Chromosomes

Culture Condition	Chromosome No.	Band	Symbol*
Rare fragile sites			
Folic acid deficiency or	2	q11.2	FRA2A
treatment with antifolates		q13	FRA2B
(Group 1)		q22.3	FRA2K
	6	p23	FRA6A
	7	p11.2	FRA7A
	8	p22.3	FRA8A
	9	p21.1[†]	FRA9A
		q32[†]	FRA9B
	10	q23.3[‡]	FRA10A
		q24.2[‡]	FRA10A
	11	q13.3[†]	FRA11A
		q23.3[†]	FRA11B
	12	q13.1	FRA12A
		q24.13[†]	FRA12D
	16	p12.3	FRA16A
	19	p13	FRA19B
	20	p11.23	FRA20A
	22	q13	FRA22A
	X	q27.3[†]	FRAXA
Distamycin A treatment	8	q24.1[†]	FRA8E
(Group 2)	16	q22.1[†]	FRA16B
	17	p12	FRA17A
BrdU treatment	10	q25.2[†]	FRA10B
(Group 3)	12	q24.2[†]	FRA12C
Unclassified	8	q13	FRA8F
(Group 4)			
Common fragile sites			
Aphidicolin treatment	1	p36	FRA1A
(Group 1)		p32	FRA1B
		p31.2	FRA1C
		p22	FRA1D
		p21.2	FRA1E
		q21	FRA1F
		q25.1	FRA1G
		q44.1	FRA1I
	2	p24.2	FRA2C
		p16.2	FRA2D
		p13	FRA2E
		q21.3	FRA2F
		q31	FRA2G
		q32.1	FRA2H
		q33	FRA2I
		q37.3	FRA2J
	3	p24.2	FRA3A
		p14.2	FRA3B
		q23	FRA3C
	4	p16.1	FRA4A
		p15	FRA4D
		q31	FRA4C
	5	q15[†]	FRA5D
		q31.1	FRA5C
	6	p25.1	FRA6B
		p22.2	FRA6C
		q13	FRA6D
		q21	FRA6F
		q26	FRA6E

continued

Table 4.2.1. continued

Culture Condition	Chromosome No.	Band	Symbol*
Common fragile sites (continued)			
Aphidicolin treatment	7	p22	FRA7B
(Group 1)		p14.2	FRA7C
		p13	FRA7D
		q21.2	FRA7E
		q22	FRA7F
		q31.2	FRA7S
		q32.3	FRA7H
		q36	FRA7I
	8	q22.1	FRA8B
		q24.1†	FRA8C
		q24.3	FRA8D
	9	q22.1	FRA9D
		q32†	FRA9E
	10	q22.1	FRA10D
		q25.2†	FRA10E
		q26	FRA10F
	11	p15.1	FRA11C
		14.2	FRA11D
		p13	FRA11E
		q13†	FRA11H
		q14.2	FRA11F
		q23.3†	FRA11G
	12	q21.3	FRA12B
		q24	FRA12E
	13	q13.2	FRA13A
		q21.2†	FRA13C
	14	q23	FRA14B
		q24.1	FRA14C
	15	q22	FRA15A
	16	q22.1†	FRA16C
		q23.2	FRA16D
	17	23.1	FRA17B
	18	q12.2	FRA18A
		q21.3	FRA18B
	19	q13	FRA19A
	20	p12.2	FRA20B
	22	q12.2	FRA22B
	X	p22.31	FRAXB
		q22.1	FRAXC
		q27.3†	FRAXD
5-AzaC treatment	1	q12	FRA1J
(Group 2)		q42	FRA1H
	9	q12	FRA9F
BrdU treatment	4	q12	FRA4B
(Group 3)	5	p13	FRA5A
		q15†	FRA5B
	9	p21†	FRA9C
	10	q21	FRA10C
	13	q21†	FRA13B
Unclassified	4	q27	FRA4E
(Group 4)	Y	q12	FRAYA

*Symbols assigned at Eighth International Workshop on Human Gene Mapping (Berger et al, 1985).

†Bands with possible coincident fragile sites.

‡FRA10A may be in 10q23.3 or 10q24.2 (After Hecht, 1988b).

folate and thymidine and to the addition of antifolates such as MTX and FUdR to culture medium (Sutherland, 1979; Glover, 1981; Tommerup et al, 1981). Folate pathways are responsible for generating products that are essential for purine and deoxythymidine-5'-phosphate (dTMP) synthesis. Conversely, MTX and FUdR promote the expression of fra(X) through inhibition of thymidylate synthetase. MTX can be converted by cells to diglutamyl MTX, which in turn inhibits thymidylate synthetase, whereas FUdR is phosphorylated in cells by thymidine kinase to form fluorodeoxyuridine monophosphate (FUdR-5'-P), an irreversible inhibitor of thymidylate synthetase. Therefore, expression of fra(X) appears to be due to depletion of dTMP levels. Various hypotheses have been proposed to explain the genetic cause of fra(X) (Freedman and Howard-Peebles, 1986; Hoegerman and Ray, 1986; Nussbaum et al, 1986; Pembrey et al, 1985).

4.22 OTHER FRAGILE SITES

At least 104 fragile sites (including those in the X chromosome) are thus far reported in human chromosomes, including 12 coincident locations expressed under different culture conditions and three unclassified fragile sites. These fragile sites are expressed in response to folate deficiency (or using treatments with antifolates) or to other chemical treatments such as aphidicolin, azacytidine (5AzaC), distamycin A (DA), and bromodeoxyuridine (BrdU). Twenty-four of them are heritable (rare) fragile sites and the remaining 80 are common fragile sites (Table 4.2.1). Unlike rare fra(X) (q27.3), the other fragile sites in the X chromosome or autosomes are so far not associated with any clinical condition. There are, however, speculations and conflicting views on their relationship as possible predisposing factors for chromosome breakpoints (cbp) in cancer-related chromosome rearrangements (DeBraekeleer et al, 1985; Hecht, 1988a; Le Beau, 1988; Sutherland, 1988a,b; Sutherland and Hecht, 1985; Sutherland and Mattei, 1987). Therefore, at least for the present, information on other fragile sites may not have direct bearing on the clinical status of the patient, but may be of interest in understanding the basic characteristics of human chromosomes and further investigations. Table 4.2.2 shows a variety of culture conditions that induce fragile sites in human chromosomes. They would be of some help in developing protocols of choice.

4.3 SISTER CHROMATID DIFFERENTIATION

Sister chromatid exchanges (SCEs) in somatic cells can be demonstrated by differentially staining each chromatid of mitotic chromosomes. This is achieved by incorporating a thymidine analog, bromodeoxyuridine (BrdU), into replicating DNA for two successive cell cycles and subsequently subjecting it to photodegradation. This results in sister chromatid differentiation (SCD) by a faintly stained chromatid containing bifiliarly BrdU-substituted DNA and

Table 4.2.2. Culture Conditions Used to Induce Rare and Common Fragile Sites

Treatment	Concentration	Duration of Treatment	References
Without folic acid		Through the entire culture period	Sutherland et al, 1983
FUdR	0.1 uM	Final 24 h	Glover, 1981; Tommerup et al, 1981
Distamycin A	100 μg/ml	Final 48 h	Schmid et al, 1980
BrdU	10 mg/L	Final 6 to 12 h	Scheres and Hustinx, 1980
	50 mg/L	Final 6 to 12 h	Croci, 1980; Sutherland et al, 1983b
Aphidicolin	0.2%	Final 26 h	Glover et al, 1984
5-Azacytidine	10^{-5} M	Final 7 h	Schmid et al, 1985

a brightly stained sister chromatid containing unifiliarly BrdU-substituted DNA (see mechanism).

This technique basically involves two distinct steps: (1) incorporation of BrdU into replicating DNA for two cell cycles, and (2) staining of chromosomes to highlight the difference in BrdU content of DNA between the two chromatids by using either fluorochrome or Giemsa.

Protocol 4.31.1   Bromodeoxyuridine (BrdU) Incorporation

Solutions

1. BrdU solution

BrdU	10 mg
Fluorodeoxyuridine (FUdR)	0.05 mg
Distilled water	10 ml

FUdR may be prepared separately at a higher concentration for convenience and added to give the required concentration in BrdU solution (dissolve 1 mg of FUdR in 10 ml of distilled water and add 0.5 ml of this solution to 9.5 ml of distilled water to which BrdU will be added).

BrdU solution can be used as such or filter sterilized before use. It should be stored frozen in the dark. Solution can be used for 2 to 3 weeks.

Procedure

1. Set up blood lymphocyte cultures as usual, following the routine protocol.
2. Incubate the cultures for 24 hours at 37°C.
3. Add 0.1 ml of BrdU solution to each tube containing 10 ml of medium and incubate the cultures in the dark for an additional 40 to 48 hours.
4. At the end of the incubation period, harvest the cultures with colcemid (0.1 μg/ml) for a final 1 hour and prepare the slides according to routine procedures.

Protocol 4.32.1   Differential Staining by Hoechst 33258

(Following Latt, 1974)

Solutions

1. Hoechst 33258 stock solution (150 μg/ml)

Hoechst 33258	1.5 mg
Distilled water	10 ml

Store frozen in dark vials. Solution can be used up to a few months.

2. Hoechst 33258 staining solution (0.5 μg/ml)

Hoechst 33258 stock solution	0.15 ml
Distilled water	45 ml

Store in a dark bottle or jar in the refrigerator. Solution can be used up to 1 month.

3. Mounting solution (0.2 M sodium phosphate, pH 7.5)

Sodium phosphate dibasic (Na$_2$HPO$_4$)	2.8 g
Distilled water	100 ml

Instead of sodium phosphate dibasic (Na$_2$HPO$_4$), either 5.4 g of sodium phosphate dibasic, 7 hydrate (Na$_2$HPO$_4$.7H$_2$O) or 7.2 g of sodium phosphate dibasic, 12 hydrate (Na$_2$HPO$_4$.12H$_2$O) can be used to prepare the same amount of solution.

Adjust the pH of the solution to 7.5 with 0.1N sodium hydroxide using pH meter if necessary. This solution can be stored cold (2° to 5°C) for 3 to 4 weeks.

Procedure

1. Stain the slides in Hoechst 33258 staining solution (0.5 μg/ml) in the dark for 10 minutes.
2. Rinse in distilled water and allow to dry.
3. Mount in mounting solution and examine using a wavelength of 360 to 400 nm.

Comments

pH of the mounting solution is critical for differential fluorescence of sister chromatids and should be at least 7.0 or more. The acidic pH will eliminate the differential fluorescence of Hoechst 33258 by interfering in the quenching process. Photography has to be performed rather quickly because Hoechst 33258 fluorescence fades rapidly.

Protocol 4.32.2   Differential Staining by Giemsa

(Following Perry and Wolff, 1974)

Solutions

1. Hoechst 33258 solution (150 μg/ml)
   Same as "Hoechst 33258 stock solution" in the above protocol.
2. 2 × saline sodium citrate (2 × SSC) solution

Sodium chloride (NaCl)	17.5 g
Sodium citrate, 2 hydrate ($Na_3C_6H_5O_7.2H_2O$)	8.8 g
Distilled water	1 L

   Adjust pH to 7.0 using 1 N sodium hydroxide solution (1N NaOH) or 1 N hydrochloric acid (1N HCl). Solution can be stored at room temperature for 2 to 3 weeks.
3. Sorensen's buffer (pH 7.0)

Potassium phosphate monobasic (KH_2PO_4)	5.26 g
Sodium phosphate dibasic (Na_2HPO_4)	8.65 g
Distilled water	1 L

   Instead of sodium phosphate dibasic (Na_2HPO_4), either 16.34 g of sodium phosphate dibasic, 7 hydrate ($Na_2HPO_4,7H_2O$), or 21.84 g of sodium phosphate dibasic, 12 hydrate ($Na_2HPO_4,12H_2O$) can be used to prepare the same amount of buffer.
4. Giemsa staining solution (2%)

Sorensen's buffer (pH 7.0)	49 ml
Gurr's Giemsa stain	1 ml

Procedure

1. Place a few drops of Hoechst 33258 solution (150 μg/ml) on each slide and cover with a cover glass. Let stand for 10 to 15 minutes in the dark at room temperature.
2. Rinse the slides in distilled water and allow to dry.
3. Mount the slides with 2 × SSC (pH 7.0) and expose to ultraviolet light at a distance of 20 to 25 cm from the source for 1 to 2 hours.
4. Rinse the slides in distilled water and incubate in 2 × SSC at 60° to 65°C for 1 to 2 hours.
5. Rinse the slides in distilled water and allow to dry.
6. Stain the slides in freshly prepared Giemsa staining solution (2%) for 15 to 20 minutes. It is preferred to monitor the staining intensity under the microscope to achieve the best differentiation.
7. Mount the slide with permount and examine the preparations using either bright-field or phase-contrast optics.

Comments

It is essential to disintegrate BrdU during the photodegradation process to obtain adequate differential staining. Prolonged exposure of preparations to radiation, however, will lead to loss of chromosome morphology. It is recommended to test run a slide or two to check the appropriate time for each batch of slides. It is possible to obtain satisfactory differential staining without incubating in 2 × SSC at 60° to 65°C.

Protocol 4.32.3   Differential Staining by DAPI

(Following Lin and Alfi, 1976)

Solutions

1. McIlvaine's buffer (pH 7.0)
   Same as in protocol 3.31.1.
2. DAPI working solution
   Same as in protocol 3.31.1.
3. Mounting solution (0.2 M sodium phosphate, pH 11.0)

   Sodium phosphate dibasic (Na_2HPO_4)   2.8 g

   Instead of sodium phosphate dibasic (Na_2HPO_4), either 5.4 g of sodium phosphate dibasic, 7 hydrate ($Na_2HPO_4.7H_2O$) or 7.2 g of sodium phosphate dibasic, 12 hydrate ($Na_2HPO_4.12H_2O$) can be used to prepare the same amount of solution.
   Adjust the pH of the solution to 10.5 to 11.0 with 0.1 N sodium hydroxide using pH meter. This solution can be stored in cold (2° to 5°C) for 3 to 4 weeks.

Procedure

1. Place a few drops of DAPI staining solution (1 μg/ml) on each slide and cover with coverslip. Stain for 10 minutes in the dark at room temperature.
2. Rinse in distilled water and allow to dry.
3. Mount in mounting solution (0.2 M sodium phosphate, pH 11.0) and blot off excess mounting solution. Examine using a wavelength in the range of 360 to 390 nm. Photography must be performed quickly, because fluorescence fades rapidly.

Comments

The differential fluorescence in microscopic examination is significant and adequately clear. The SCE pattern could be observed without any prior treatment. Although the fluorescence pattern seen by DAPI is intense and esthetically beautiful, the photographic reproduction is rarely satisfactory. Resolution under direct microscopic observation is far superior to that which could be reproduced. Another disadvantage of performing SCE using either fluorochrome is the rapidity with which fluorescence fades. It is therefore difficult to examine a metaphase for a sufficiently long period. Any attempt to stabilize fluorescence by adding glycerol to mounting solution interferes in quenching fluorescence and totally eliminates SCD.

Staining Pattern

Sister chromatid differentiation is observed in metaphases obtained from cells that have incorporated BrdU into DNA at least for two consecutive cell cycles (Figure 4.3.1A and B). Cells that have replicated for one cell cycle in presence of BrdU do not exhibit SCD. Furthermore the cells that replicated more than two cell cycles do not show SCD in all chromosomes, but may have (a) chromosomes with both faintly stained chromatids, (b) chromosomes with a dark and a faint chromatid (similar to that of M-2), or (c) chromosomes with a portion that has SCD and the remaining portion that has both lightly stained chromatids (Figure 4.3.1C

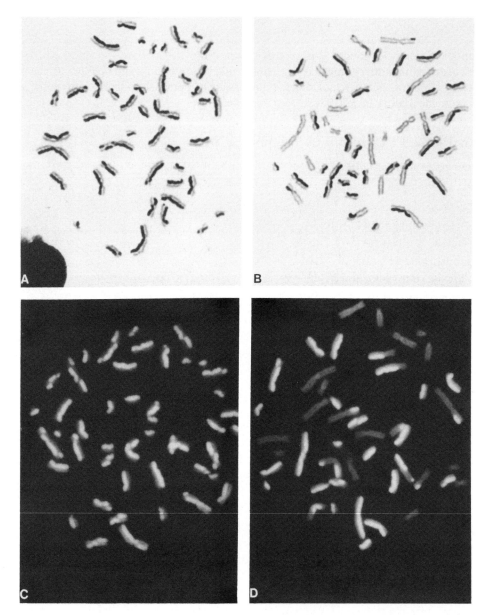

FIGURE 4.3.1. Metaphases showing sister chromatid differentiation by Giemsa (A, B) and DAPI (C, D). Cells shown in A and C have undergone replication for two cell cycles (M_2), whereas those in B and D have completed more than two replication cycles (M_3).

and D). SCD can be visualized either by using one of the fluorochromes, Hoechst 33258 or DAPI, or by using Giemsa subsequent to photodegradation of BrdU. The chromatid containing DNA with one BrdU-incorporated strand stains more intensely than does the chromatid containing DNA with both BrdU-incorporated strands. Sister chromatid exchanges are analyzed by counting the number of exchanges as indicated by discontinuity of intensely stained chromatid in each cell. Each point of discontinuous staining is enumerated as an exchange (Figure 4.3.2). Larger chromosomes usually have a greater number of SCE than do smaller ones. A chromosome may not have any SCE, whereas its homologue may show one or more SCE in a given cell. Likewise, the incidence may vary from cell to cell. Although the SCD

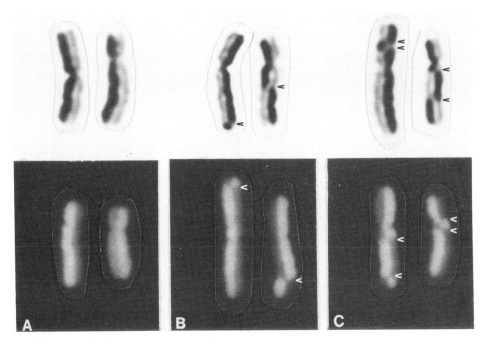

FIGURE 4.3.2. Selected chromosome stained by Giemsa (*top row*) and DAPI (*bottom row*) showing no sister chromatid exchange (SCE) (A), one SCE (B), and two SCEs (C).

is adequate with any of the procedures just described, the SCE frequency revealed by photodegradation plus Giemsa reportedly is consistently higher than that of fluorochrome Hoechst 33258 (Perry and Wolff, 1974).

Mechanism

To obtain SCD, asymmetric incorporation of BrdU in chromosomal DNA is essential. The asymmetry could be either one unifiliarly BrdU-incorporated chromatid and the other bifiliarly BrdU-incorporated chromatid (ie, cells grown consecutively for two cell cycles in the presence of BrdU) or one chromatid containing a single BrdU-substituted strand and the other containing unsubstituted DNA (ie, cells grown in the presence of BrdU for one cell cycle and subsequently grown for another cell cycle without BrdU) (Figure 4.3.3). Two basic mechanisms have been suggested for SCD staining: (1) the procedures used for differential staining causes excessive loss of DNA from a bifiliarly BrdU-substituted chromatid than does its counterpart and the staining intensity is proportional to the amount of DNA remaining in the chromatid, and (2) the alteration in chromosomal proteins brought about by BrdU incorporation into DNA may in some way influence the binding of Giemsa stain to chromosomes, resulting in differential staining.

Applications

Sister chromatid exchanges are seen in most normal eukaryotic cells including human. The number of SCE in normal human cells varies in different cells, ranging from 2 to 20 with a mean frequency of SCE between 5 and 8 (Carrano et al, 1980). The mean number of SCE does not appear to differ between sexes, but it may be influenced by age. The maximum frequency of SCE is observed between the ages of 30 and 40 years (de Arce, 1981). It is uncertain if this baseline SCE reflects the naturally occurring exchanges or the intrinsic effect of BrdU incorporation into DNA.

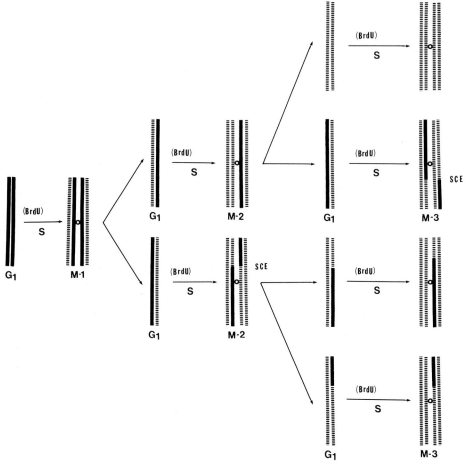

FIGURE 4.3.3. Diagrammatic illustration of mechanism of sister chromatid exchanges. *Solid bars* indicate the native DNA strand, whereas *hatched bars* indicate BrdU incorporated DNA strand.

The technologic invention of SCD using BrdU has been far superior in resolution to score SCE and technically much simpler to follow than the earlier methods using radioactive nucleotides. Soon after inception of the new technique, it was used extensively in testing the mutagenic potential of several known as well as new chemicals by comparing the SCE frequency induced by test agent and the normal incidence (Gebhart, 1981). Although a number of known clastogens produce increased frequency of SCE, a positive correlation is observed only in about 70 percent of test agents, whereas no such relation is evident in the remaining. This lack of consistent relationship between the two parameters is interesting, that is, certain test agents elicited SCE but not structural chromosome aberrations and vice versa. From these studies it is evident that SCE and structural chromosome aberrations are different and independent expressions of mutagen-induced damage and probably signify different end results. For example, cells with structural aberrations most likely proceed towards senescence, whereas those with increased SCE are compatible with survival. The increased SCE may eventually lead to a higher rate of mutagenesis. In other words, SCE are the primary visible events that possibly reflect the long-term effect of mutagens. Regardless of the limitations of the SCE test system, in light of the many "false-positive" and "false-negative" results and the difficulty in interpretation of such results due to the lack of adequate understanding of biologic mechanisms of SCE, they can be an important additional parameter that can provide important information on the cytogenetic effects of mutagens.

Sister chromatid differentiation can be used effectively to study cell kinetics, for example, in determining duration various stages and the total cell cycle (Craig-Holmes and Shaw, 1976; Crossen and Morgan, 1977). In addition, the BrdU labeling method is applicable in distinguishing first, second, and third generation mitosis whenever such delineation is required.

Among chromosome breakage syndromes, a significant increase in the number of SCE compared with the baseline SCE in normal human cells is observed in Bloom's syndrome (German et al, 1974). There is no increase in SCE in other chromosome breakage disorders, such as Fanconi's anemia, ataxia telangiectasia, and xeroderma pigmentosum. Furthermore, increased SCE frequency is observed when the cells of patients with xeroderma pigmentosum are exposed to either ultraviolet radiation or to a carcinogen 4-nitroquinoline-1-oxide (San et al, 1977; Wolff et al, 1977).

4.4 REPLICATION PATTERNS

Elaborate techniques involving incorporation of tritiated nucleotides, thymidine in particular, and autoradiography have become virtually obsolete in investigating DNA replication in metaphase chromosomes with the inception of bromodeoxyuridine (BrdU). BrdU is a thymidine analog that can be incorporated into the replicating DNA in place of thymidine. The photodegradable property of BrdU has enabled the microscopic demonstration of chromosomal regions consisting of BrdU-substituted DNA by deferential staining. The BrdU labeling of replicating DNA has overwhelming advantages such as technical simplicity and superb resolution.

4.41 BrdU LABELING

Various replication events of chromosomes can be studied by labeling the DNA for a particular time period. They include two types of labeling: (1) continuous labeling, and (2) discontinuous or pulse labeling. In general practice, continuous labeling is often used to investigate replication events that occur at the end of the synthetic phase (S phase) or, in other words, late replicating patterns. Discontinuous labeling instead is used to reveal the chromosomal regions that undergo replication during early and mid-phases of the S period.

Protocol 4.41.1 Continuous Labeling

(Following Dutrillaux et al, 1973; Latt, 1973)

Solutions

1. Fluorodeoxyuridine (FUdR) stock solution (4×10^{-4}M)

FUdR	1 mg
Distilled water	10 ml

   Store frozen in dark vials.

2. Uridine stock solution (3×10^{-3}M)

Uridine	1 mg
Distilled water	1 ml

   Store frozen as small aliquots.

3. BrdU solution

BrdU	3 mg	$(10^{-2}$ M)
FUdR stock solution $(4 \times 10^{-4}M)$	0.1 ml	$(4 \times 10^{-5}$ M)
Uridine stock solution $(3 \times 10^{-3}$ M)	0.2 ml	$(6 \times 10^{-4}$ M)
Distilled water	0.7 ml	

Store frozen in a dark vial. Filter sterilization is generally not required. The solution can be used over a period of 2 weeks.

Procedure

1. Set up the lymphocyte cultures following routine protocols. Incubate the cultures for about 60 to 65 hours at 37°C.
2. Add 0.1 ml of BrdU solution to each 10 ml of culture medium. (Final concentrations of BrdU, FUdR, and uridine in the medium are 10^{-4}M, 4×10^{-7} M, and 6×10^{-6} M, respectively.)
3. Incubate the cultures in dark (by wrapping with black paper or aluminum foil) for 5 to 6 hours.
4. Add 0.02 ml of colcemid and incubate the cultures for an additional 1 hour in the dark.
5. Harvest the cultures like regular lymphocyte cultures.
6. Slides are stained by one of the staining procedures to be described in this section.

Comments

Cultures should be incubated in the dark after adding BrdU solution to prevent photo-degradation of BrdU. Chromosomes of BrdU-incorporated cells tend to remain relatively longer than do untreated cells. Slightly longer hypotonic treatment would help to obtain better chromosome spreads. The same protocol can be used for monolayer cultures without any modifications.

BrdU incorporation could be extended for much longer periods to obtain a variety of bands. Various patterns of replications by continuous labeling have been described into five different stages (I to V), depending on the extent of BrdU labeling (Camargo and Cervenka, 1980).

Replication Pattern

BrdU labeling during the final hours of culture give late replication patterns. The patterns obtained by either staining method are similar to R-bands. Heavy staining of the R-banded regions indicate that they have very little, if any, BrdU labeling because they have completed their replication early in the S phase before BrdU was added to the cultures. The lightly stained regions correspond to the G-bands or Q-bands which have much of the incorporated BrdU. It is well documented that the G-bands undergo replication towards the end of the S phase. The large heterochromatic segments (secondary constriction or h regions) in chromo-somes 1, 9, and 16 are heavily labeled. In males, the heterochromatic distal long arm region of the Y shows a significant amount of BrdU incorporation. In females, the inactive or het-erochromatic X (lionized X) chromosome replicates late and is excessively labeled with BrdU compared with the rest of the complement and its homolog, the functional X chromosome.

Applications

Late replication profiles are being used as R-bands (RBG, R-bands by BrdU using Giemsa, and RBA, R-bands by BrdU using acridine orange techniques) in a number of studies includ-ing routine clinical evaluation. Chromosomes obtained by BrdU incorporation tend to remain relatively long and thus provide greater resolution than do regular preparations. The replication sequence in the female inactive X chromosome was demonstrated to be heterogeneous and does not follow a uniform pattern (Babu et al, 1986; Willard, 1977; Willard and Latt, 1976). BrdU incorporation during the late replicating phase has been adopted to obtain high resolution R-bands after the release of mitotic block in synchronization procedures (Pai and Thomas, 1980).

Protocol 4.41.2   Discontinuous Labeling

(Following Dutrillaux et al, 1976)

Solutions

1. BrdU solution
    Same as in protocol 4.41.1.
2. Thymidine solution (1×10^{-3} M)

Thymidine	2.5 mg
Distilled water	10 ml

   Filter sterilize and store frozen in small aliquots. Solution is relatively stable and can be used for 2 to 3 months.

Procedure

1. Set up the lymphocyte cultures according to routine protocols.
2. Incubate the cultures at 37°C for 50 to 55 hours.
3. Add 0.1 ml of BrdU solution to each 10 ml of medium. Continue the incubation of cultures for 10 to 15 hours (total incubation period 60 to 65 hours) in the dark. (Wrap the cultures with either black paper or aluminum foil.)
4. At 60 to 65 hours, centrifuge the cultures at 800 rpm for 10 minutes. Discard the supernatant and wash the cells once with regular medium (without exposing the cultures to bright lights).
5. Resuspend the cells in the same amount of regular culture medium and add 0.1 ml of thymidine solution (10^{-3} M).
6. Incubate the cultures in the dark for an additional 6 to 7 hours with the addition of 0.02 ml of colcemid for the final 1 hour.
7. Harvest the cultures by the usual procedures.
8. Slides are stained by either acridine orange or Giemsa protocol to be described.

Comments

Cultures should be prevented from exposure to bright light during the washing and addition of renewal medium. Cultures should be kept in the dark from the time of BrdU addition until the harvest. See also the comments in protocol 4.41.1.

Replication Patterns

A wide variety of replication profiles can be seen with discontinuous BrdU labeling. Some metaphases are seen with typical staining patterns similar to those of G- or Q-bands. This phenomenon is opposite to that seen in continuous labeling for late replication patterns. The darkly stained bands have little incorporation of BrdU, reflecting that they are not replicating in the early part of the S phase. The major heterochromatic regions in chromosomes 1, 9, and 16 do not show any labeling and are darkly stained in this type of cell. Similarly, the heterochromatic region of the Y in males and the inactive X chromosome in females remain darkly stained. Other cells from the same preparations demonstrate a variety of bands that are in intermediate stages between G- and R-bands. These bands represent the events of replication during mid-S phase (Camargo and Cervenka, 1980; Dutrillaux et al, 1976; Kaluzewski, 1982).

4.42   STAINING OF BrdU-LABELED CHROMOSOMES

Bromodeoxyuridine-labeled chromosomes can be stained by a fluorescent dye, Hoechst 33258 or acridine orange, or by Giemsa stain after the photodegradation process of BrdU. Among the three staining methods, acridine orange and Giemsa stain have become the choice of many investigators primarily because of greater differentiation and better resolution.

Protocol 4.42.1   Acridine Orange Staining

Solutions

1. Sorensen's buffer (0.06 M, pH 6.5)

Potassium phosphate monobasic (KH_2PO_4)	2.51 g
Sodium phosphate dibasic (Na_2HPO_4)	5.87 g
Distilled water	1 L

Instead of sodium phosphate dibasic (Na_2HPO_4), 11.09 g of sodium phosphate dibasic, 7 hydrate ($Na_2HPO_4 \cdot 7H_2O$), or 14.82 g of sodium phosphate dibasic, 12 hydrate ($Na_2HPO_4 \cdot 12H_2O$), can be used to prepare the same amount of buffer.

2. Acridine orange staining solution (0.01%)

Acridine orange	10 mg
Sorensen's buffer (0.6 M, pH 6.5)	100 ml

Store in the dark at 2° to 5°C. The solution can be used for 1 to 2 months.

Procedure

1. Stain the slide in acridine orange staining solution (0.01%) for 5 minutes in the dark.
2. Rinse the slide briefly in Sorensen's buffer (pH 6.5).
3. Mount the slide with the same buffer and blot the excess fluid.
4. Examine using a fluorescence microscope with filter combination for 420 to 460 nm.

Comments

The extent of acridine orange staining is crucial in obtaining proper differentiation and resolution. The overstained preparations show overall bright red fluorescence. The excess stain can be removed by lifting off the cover glass and rinsing in buffer (just one or two dips). The slides can be processed until optimum staining is achieved. Conversely, if the preparations are understained, as recognized by too much green color, they can be restained by replacing them in staining solution. The right stage of staining is that which appears light reddish over the cells and just turns green after exposure to the ultraviolet source. Chromosomes with little or no cytoplasm surrounding them show better staining, whereas those with too much cytoplasm do not. Fluorescence patterns can be photographed with either a color film (Kodachrome 64) or a panchromatic black and white film.

Staining Pattern

Acridine orange gives green fluorescence when bound to double-stranded DNA and red fluorescence with single-stranded DNA and RNA. The chromosomal regions that show green fluorescence, therefore, have natural double-stranded DNA. Contrary to this, regions with red fluorescence have single-stranded DNA due to photodegradation of incorporated BrdU. The chromosome regions with green fluorescence correspond to the dark bands by Giemsa staining. (See next protocol.)

Protocol 4.42.2   Giemsa Staining

(Modified from Perry and Wolff, 1974)
    Same as in protocol 4.32.2.

Staining Pattern
The chromosome regions that replicated in the presence of BrdU consist of BrdU-substituted DNA. These regions are faintly stained, whereas those that have DNA with natural bases are darkly stained. Giemsa staining provides an advantage over acridine orange staining because of the permanent nature of the preparations (Figure 4.3.4).

4.5   LATERAL ASYMMETRY

Chromosomes from cells grown in the presence of BrdU for two consecutive cell cycles show asymmetrical staining between two chromatids when appropriately stained. However, cells grown for only one replication cycle, when stained with either fluorochrome Hoechst 33258 or DAPI, exhibit asymmetrical fluorescence within the C-band regions of some chromosomes. Lateral asymmetry in certain chromosome regions can further be visualized using Giemsa stain.

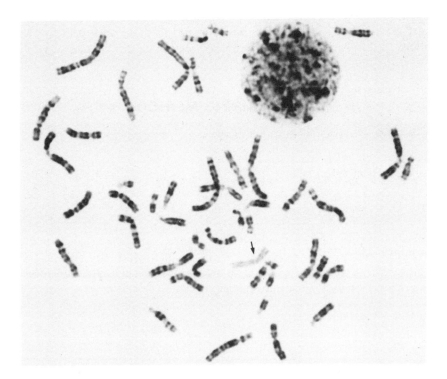

FIGURE 4.3.4. Metaphase stained by RBG technique from a cell cultured in the presence of BrdU for 5.5 hours before harvesting. The late replicating X chromosome is marked by an arrow.

4.51   BrdU INCORPORATION

Protocol 4.51.1   Incorporation of BrdU into Replicating DNA

(Following Latt et al, 1974)

Solutions

1. BrdU solution
   Same as in protocol 4.41.1.

Procedure

1. Set up the peripheral blood lymphocyte cultures according to routine protocols.
2. Incubate cultures at 37°C for 48 or 72 hours.
3. Add 0.1 ml of BrdU solution to 10 ml of culture medium and incubate for 24 hours. (Final concentration of BrdU is 20 μg per milliliter of culture medium.) (Alternatively, BrdU may be incorporated into the culture medium at the initiation of the cultures, and the cultures can be harvested at 45 to 50 hours.)
4. Add 0.02 ml of colcemid (10 μg/ml) to each culture and incubate for an additional 60 minutes.
5. Harvest the cultures and prepare slides following routine procedures described in the protocols.
6. Stain the slides according to one of the following methods.

Comments
The suggested final concentrations of BrdU in culture medium for incorporation vary greatly from about 30 to 40 μg/ml (Latt et al, 1974; Kim, 1975). However, if lower concentrations of BrdU are preferred, it is advisable to introduce FUdR along with BrdU in the cultures. FUdR inhibits the synthesis of deoxythymidine-5'-phosphate (DTMP) and favors the incorporation of BrdU into replicating DNA.

4.52   STAINING METHODS

Protocol 4.52.1   Using Hoechst 33258

(Following Latt et al, 1974; Lin et al, 1974)
   Same as in protocol 4.32.1.

Protocol 4.52.2. Using DAPI

(Following Lin and Alfi, 1976; Lin et al, 1977)
   Same as in protocol 4.32.3.

Protocol 4.52.3   Using Alkaline Giemsa Stain

(Following Brito-Babapulle, 1981)

Solutions

1. Sodium phosphate solution (0.3 M, pH 10.4)

Sodium phosphate dibasic (Na_2HPO_4)	4.26 g
Distilled water	100 ml

Instead of sodium phosphate dibasic (Na$_2$HPO$_4$), 8.04 g of sodium phosphate dibasic, 7 hydrate (Na$_2$HPO$_4$·7H$_2$O) or 10.74 g of sodium phosphate dibasic, 12 hydrate (Na$_2$HPO$_4$·12H$_2$O) can be used to prepare the same amount of solution.

Adjust pH to 10.4 using 0.1 M sodium hydroxide (NaOH).

2. Giemsa staining solution (2%)

Sodium phosphate solution (pH 10.4)	50 ml
Gurr's Giemsa stain	1 ml

Procedure

1. Age the prepared slides according to protocol 4.51.1 for 1 or 2 days.

2. Stain the slides in Giemsa staining solution (2%) for 10 to 12 minutes at room temperature. (It is suggested that the staining intensity be monitored under the microscope during the staining process.)

3. Rinse in deionized water and allow to dry.

4. Mount the slide using permount and examine with either bright-field or phase-contrast microscopy.

Protocol 4.52.4   Using Photodegradation and Giemsa Stain

(Modified from Angell and Jacobs, 1975; Perry and Wolff, 1974)

Same as in protocol 4.32.2.

Staining Patterns

The sister chromatids in the peracentric regions of some chromosomes show staining differences (lateral asymmetry) (Figure 4.3.5) Lateral asymmetry has been reported in the peracentric regions of chromosomes 1, 2, 3, 4, 5, 7, 9, 13, 14, 15, 16, 17, 20, 21, and 22, and the long arm of Y (Latt et al, 1974; Kim, 1975; Brito-Babapulle, 1981; Ghosh et al, 1979; Galloway and Evans, 1975). Some chromosomes, such as 1, 3, 4, 15, 16, and the Y, show lateral asymmetry more frequently than do others (Brito-Babapulle, 1981; Galloway and Evans, 1975).

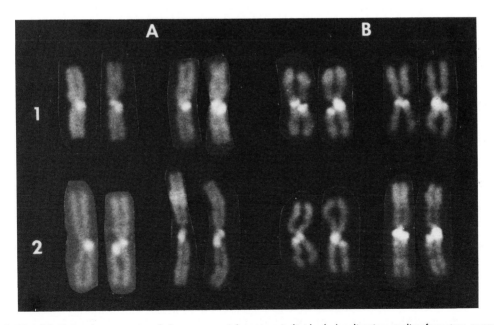

FIGURE 4.3.5. Lateral asymmetry of chromosome 1 from two individuals (replication studies from two experiments). (Courtesy of Dr. M. S. Lin.)

Mechanism

The lateral asymmetry in staining intensity between specific regions of two chromatids of the same chromosome reflects the unequal distribution of the thymine-rich chain. Initial findings on the C-bands of mouse chromosomes led to the belief that the satellite DNA, which contains a strand with 45 percent thymine and the other with 22%, is responsible for an asymmetric staining pattern (Flamm et al, 1967). A similar relation was suggested in human chromosomes between the regions exhibiting lateral asymmetry and those containing satellite DNAs (Galloway and Evans, 1975). Subsequent studies on human chromosomes, however, revealed an asymmetric pattern in other chromosomal regions that have no significant amount of satellite DNA sequences, suggesting that other types of repetitious DNA may exhibit this phenomenon as well and may contribute to asymmetry.

The unequal distribution of thymine in DNA leads to unequal incorporation of BrdU into replicating DNA. Following a single replication cycle, the DNA in one chromatid receiving a thymine-rich strand will have much less BrdU than will the corresponding region of the sister chromatid that has DNA rich in guanine. The unequal amounts of BrdU are responsible for asymmetric staining between regions of chromatids.

Asymmetry could be either simple or compound. In simple asymmetry, the entire C-band region in one chromatid is more intensely stained than the other. This pattern is supposedly due to thymine bias in a single strand of DNA of the whole C-band. However, different blocks of the same C-band have unequal staining in compound asymmetry that apparently reflects thymine bias only in certain stretches of DNA within the region.

Applications

The centromeric C-bands of human chromosomes 1, 3, 4, 15 to 17, and 19 to 22, and the C-band in the long arm of Y have been described as exhibiting at least some kind of asymmetry. Lateral asymmetry has characteristics similar to those of other heteromorphisms. The asymmetric pattern in a chromosome is consistent within an individual, and it is inherited as a simple mendelian trait (Angell and Jacobs, 1978). The lateral asymmetry of certain chromosomes can therefore be applied as chromosome markers (Comings, 1978).

4.6   CHROMATIN UNDERCONDENSATION TECHNIQUES

When cells are treated with certain DNA binding agents and DNA base analogs before harvesting, some chromosome regions fail to undergo a normal condensation cycle and remain elongated. This deferential condensation is observed in metaphase chromosome preparations. DNA binding agents Hoechst 33258, DAPI, and distamycin A (DA) and DNA base analogs 5-azacytidine (5-Aza-C) and 5-azadeoxycytidine (5-Aza-dC) have been used to produce differential condensation in human chromosomes (Table 4.6.1). The optimal concentration and duration of treatment for each compound are described in the following protocols.

Protocol 4.61.1   Hoechst 33258 Treatment

(Following Pimpinelli et al, 1976)

Solutions

1. Hoechst 33258 solution (0.25%)

Hoechst 33258	25 mg
Distilled water	10 ml

Dissolve and store frozen in the dark. Filter sterilization is not required.
Add 0.1 ml/5 ml of culture medium to give a final concentration of 50 μg/ml.

Table 4.6.1. Effect of Treatment by DNA Ligands and DNA Base Substitutes on Human Chromosome Condensation

Chemical*	Optimum Concentration of Treatment	Optimum Duration of Treatment (h)	Mechanism of Induction of Undercondensation	Effected Phase of Cell Cycle	Effected Regions of Human Complement	References
Hoechst 33258[†]	50 μg/ml	7	Binding to DNA	G_2	Heterochromatin in the long arm of the Y	Marcus et al, 1979
DAPI	100 μg/ml	17	DNA intercalation	S phase	Heterochromatin in chromosomes 1, 9, 16, the Yq, and p-arms of some D- and G-groups	Prantera et al, 1981
DA	100 μg/ml	24	DNA intercalation	S phase	Heterochromatin in the Yq	Schmid et al, 1984
5-Aza-C	3.5×10^{-7} M	7–8	Substituting cytidine in replicating DNA	Late S phase	Heterochromatin in chromosomes 1, 9, 16, the Yq, and p-arms of some D- and G-groups	Schmid et al, 1983
5-Aza-dC	1×10^{-5} M	>24	Substituting cytidine in DNA for two replication cycles	Two successive S phases	One chromatid of each chromosome consisting of bifiliarly 5-aza-dC substituted DNA	Haaf et al, 1986

5-Aza-C = 5-azacytidine; 5-Aza-dC = 5-azadeoxycytidine.

*BrdU, a thymidine analog used for replication pattern studies, which also produces uneven condensation when incorporated into replicating DNA, is not included in this section.

[†]Effect of Hoechst 33258 is most profound in fibroblasts and lymphoblasts but not in lymphocytes.

Procedure

1. Add 0.1 ml of Hoechst 33258 solution to an actively growing fibroblast or lymphoblast (but not lymphocyte) culture containing 5 ml of medium and incubate it in the dark at 37°C for an additional 6 hours.
2. Harvest the cultures following routine procedures depending on the type (monolayer or suspension) of culture.
3. The slides can be stained with Giemsa. Banding procedures can be applied whenever necessary.

Comments

The effect of Hoechst 33258 on chromosome condensation is most pronounced on fibroblasts and lymphoblasts than on peripheral blood lymphocytes (Kucherlapati et al, 1975; Marcus et al, 1979a; Pimpinelli et al, 1976). It is therefore preferable to use fibroblast cultures derived from tissues such as skin or amniotic fluid. Cultures should be kept in the dark after Hoechst 33258 is added.

Undercondensation Pattern

Differential chromosome condensation by Hoechst 33258 depends on the source and type of human cells used for treatment (Marcus et al, 1979a). Chromosomes obtained from human fibroblasts treated with Hoechst 33258 show overall undercondensation, which is remarkable in the heterochromatic region of the Y chromosome. This differential undercondensation of the human Y chromosome is also observed in mouse-human hybrid cells (Marcus et al, 1979b). The mouse L-cells treated with Hoechst 33258 show profound undercondensation of centromeric heterochromatin of metaphase chromosomes (Hilwig and Gropp, 1973). A similar effect, however, is not found in primary cell cultures of mouse (Marcus et al, 1979).

Protocol 4.62.1   DAPI Treatment

(Following Prantera et al, 1981)

Solutions

1. DAPI solution

DAPI	1 mg
Culture medium	1 ml

DAPI does not dissolve readily in solutions containing phosphates. Make a fresh suspension before use and add to the cultures. Do not filter. Sterility is usually not a problem.

Procedure

1. Add 1 ml of DAPI solution to exponentially growing cell cultures (either lymphocytes or fibroblasts) containing 10 ml of culture medium.
2. Incubate in the dark for an additional period of 16 to 17 hours.
3. Harvest the cultures following routine protocols.
4. Stain the preparations as required using either for unbanded or banded chromosomes.

Comments

If peripheral blood lymphocyte cultures are being used, a few drops of heparin can be added before adding DAPI to prevent clotting that may subsequently occur. Incubate the cultures in the dark after introducing DAPI. Sterility of cultures is usually not a problem if the DAPI solution is prepared fresh and used immediately. The mitotic index may be lower than that of the normal untreated cultures.

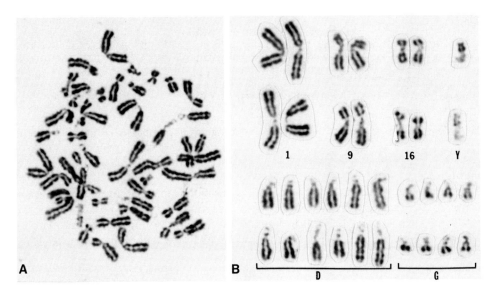

FIGURE 4.6.1. (A) A metaphase showing undercondensed chromosome regions by in vitro DAPI treatment. (B) Selected chromosomes 1, 9, 16, and the Y and D- and G-group chromosomes from two different cells showing various degrees of undercondensation.

Undercondensation Pattern

The major heterochromatic regions of chromosomes 1, 9, 16, and the Y remain distinctly undercondensed (Figure 4.6.1). The Y heterochromatin is more frequently undercondensed than is the heterochromatin in the autosomes (Prantera et al, 1981).

Protocol 4.63.1   Distamycin A (DA) Treatment

(Following Prantera et al, 1979)

Solution

1. DA solution

DA	2 mg
Culture medium	2 ml

   DA is unstable in solution. Prepare fresh before use. DA may be filter sterilized after dissolving in medium, but it is not essential.

Procedure

1. Add 1 ml of DA solution to a culture containing 10 ml of medium to give a final concentration of 100 μg/ml. Incubate the cultures in the dark at 37°C for an additional 24 hours.

2. Harvest the cultures following the usual procedures.

3. The slides may be stained with Giemsa for unbanded chromosomes or by any other method for banded chromosomes.

Comments

Incubate the cultures in the dark after adding DA solution. It may require slightly longer hypotonic treatment to obtain proper chromosome spreading.

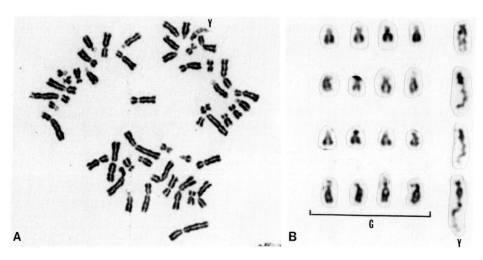

FIGURE 4.6.2. (A) A metaphase from a lymphocyte culture, treated with distamycin A for final 24 hours, showing a Y chromosome with undercondensed heterochromatic region (*marked*). (B) Selected G-group chromosomes and the Y chromosome from four different cells from the same culture, showing the extent of undercondensation in the long arm of Y chromosomes.

Undercondensation Pattern

Distamycin A treatment of human lymphocyte cultures inhibits condensation of the heterochromatin in the Y chromosome. The other major heterochromatic regions of autosomes are apparently unaffected (Schmid et al, 1984) (Figure 4.6.2).

Protocol 4.64.1   5-Azacytidine (5-Aza-C) Treatment

(Following Schmid et al, 1984)

Solutions

1. 5-Aza-C solution

    5-Aza-C stock solution (3.5×10^{-4} M)

5-Aza-C	1 mg
Distilled water	11.5 ml

    5-Aza-C working solution (3.5×10^{-5} M)

5-Aza-C stock solution	1 ml
Distilled water	9 ml

    Store frozen in the dark in small aliquots.

Procedure

1. Add 0.1 ml of 5-Aza-C working solution (3.5×10^{-5} M) to an exponentially growing culture containing 10 ml of culture medium (to give a final concentration of 3.5×10^{-7} M 5-Aza-C per milliliter).

2. Incubate the cultures in the dark for an additional 7 hours.

3. Harvest the cultures following routine procedures.

4. Stain the slides for unbanded or banded chromosomes as required.

Comments

Cultures treated with concentrations between 5×10^{-7} M and 2×10^{-7} 5-Aza-C for 7 hours produce about 80% of cells showing undercondensation of heterochromatin in chromosomes 1, 9, 15, 16, and Y (type I). Treatment with higher concentrations leads to lower mitotic indices and either numerous undercondensed sites in chromosomes that appear to be pulverized (type III) or segmentation of chromosomes similar to the R-banding pattern (type II). Conversely, treatment using lower concentrations reduces cells with differential condensation. Duration of treatment is also critical in obtaining optimal preparations. Longer treatment increases type II and type III metaphases, whereas shorter periods produce fewer metaphases showing undercondensation (Schmid et al, 1984; Viegas-Pequignot and Dutrillaux, 1976; 1981).

Undercondensation Pattern

Human lymphocytes treated with appropriate levels of 5-Aza-C show selective undercondensation of heterochromatic regions in chromosomes 1, 9, 15, 16, and the Y (Figure 4.6.3). All the chromosomal regions may not be found in undercondensed state in the same cells. In treated cultures, the heterochromatic regions of chromosomes 1 and 9 are seen undercondensed in more cells than are those of chromosomes 16 and Y (Schmid et al, 1984).

Protocol 4.65.1   5-Azadeoxycytidine (5-Aza-dC) Treatment

(Following Haaf et al, 1986)

Solutions

1. 5-Aza-dC solution (10^{-3} M)

5-Aza-dC	2 mg
Distilled water	8 ml

The solution may be filter sterilized and stored frozen in dark.

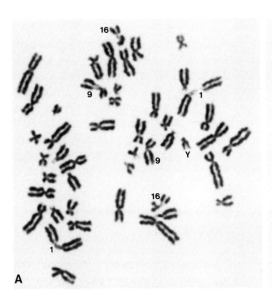
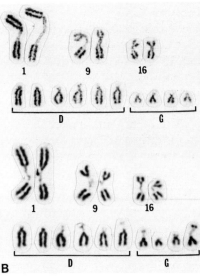

FIGURE 4.6.3. (A) A male metaphase from a lymphocyte culture, treated with 5-azacytidine for 7 hours before harvesting, showing undercondensed regions in chromosomes 1, 9, 16, and Y. (B) Selected chromosomes 1, 9, 16, and D- and G-group chromosomes from two different cells, showing undercondensation pattern.

Procedure

1. Add 0.1 ml of 5-Aza-dC solution (10^{-3} M) to a lymphocyte culture containing 10 ml of medium (to give a final concentration of 10^{-5} M 5-Aza-dC) at 48 hours.

2. Incubate the cultures for an additional 24 hours in the dark at 37°C.

3. Harvest the cultures at 72 hours following a routine protocol.

4. Stain the slides with Giemsa stain for unbanded chromosomes.

Comments

Cultures grown in 5-Aza-dC at 10^{-5} M concentration for 24 hours show differential condensation in sister chromatids in about 50 percent of the cells. Treatment with higher concentrations of 5-Aza-dC is toxic and significantly reduces mitotic index.

Undercondensation Pattern

The cells that replicated for two replication cycles in the presence of 5-Aza-dC show differential condensation between two sister chromatids. The chromatid containing DNA bifiliarly substituted with 5-Aza-dC is significantly longer than is the other containing unifiliarly substituted DNA.

Applications

Because DA treatment specifically prevents normal condensation in the heterochromatic region of the Y chromosome, it can be used to categorically demonstrate involvement of the Y in some of the translocations (Schmid, 1979). Identification of the Y chromosome with conventional straining procedures is far more difficult in a translocation between the Y and an acrocentric chromosome, because of a high degree of heteromorphism associated with the short arm of the acrocentric chromosomes. DA treatment has proved to be a very valuable tool in discriminating the Y heterochromatin from the normal heteromorphisms of acrocentric chromosomes in one such translocation (Schmid et al, 1983). Furthermore, experimental undercondensation has been used to understand the relation between the argentinofilic (silver stainable) proteins and chromatin condensation (Haaf et al, 1984).

Mechanisms

Chemicals capable of altering the normal condensation of one or more regions of chromosomes can be categorized into two basic types. Hoechst 33258, DAPI, and DA are the AT-specific DNA ligands, whereas 5-Aza-C, 5-Aza-dC, and BrdU (not described in this section) are the base analogs. The former group of chemicals bind to AT-rich double-stranded DNA and prevent normal condensation. On the contrary, the latter group of DNA base analogs must be present in the medium during replication and must be incorporated into the replicating DNA to bring about undercondensation (Haaf et al, 1984). This is the reason that cells treated for different lengths of time with these base analogs exhibit a variety of undercondensed patterns in chromosomes (Schmid et al, 1984; Viegas-Pequignot and Dutrillaux, 1976; 1981).

4.7   X- AND Y-CHROMATIN

Sex chromosome constitution of human cells can be tentatively ascertained for rapid screening without performing elaborate chromosome analysis. These techniques basically use heterogeneous staining of X-chromatin in interphase nuclei by an appropriate staining procedure and the brilliant fluorescence of the heterochromatic region in the long arm of Y by quinacrine. The routine protocols used in clinical studies are described herein. However, variations of these methods can be applied in a number of other specimens.

4.71   X-CHROMATIN

Barr and Bertram (1949) described a darkly stained chromatin body in the neurons of the female cat that is not seen in male cats. The heterochromatic body found exclusively in female cells is termed the "Barr body" or sex chromatin and is found in many mammalian species including human. It later became evident that the sex chromatin represents the inactive X chromosome or the lyonized X (Lyon, 1962) which remains in a condensed state through the interphase and manifests as a heterochromatic body. For routine clinical investigation, the buccal smear is the most easily available and common source of cells for sex chromatin analysis.

Protocol 4.7.1   Barr Body Staining

Solutions

1. Aceto-orcein stain solution (0.2%)
   Prepared stain solution is purchased and used as is.

Procedure

1. Clean the inside of the cheek with a sterile cotton swab or 2 × 2 inch gauze.
2. Gently scrape buccal membrane from both sides with a spatula.
3. Spread the material containing buccal mucosa over the slide.
4. Stain the slide with acetoorcein by placing 3 to 4 drops of stain solution (0.2%).
5. Place a coverglass over the slide and squeeze out excess stain by pressing between several layers of bibulous paper.
6. Examine the slide under the microscope. Locate the area containing sheets of epithelial cells using low power objective (10x or 16x) and analyze the cells for chromatin body under oil immersion.

Comments

Buccal smear usually contains a significant microbial population. The insides of the cheeks should be cleaned before obtaining the sample of buccal smear. Slides are examined immediately after the preparation.

Staining Pattern

Sex chromatin is seen as a darkly stained body usually at the periphery of the nucleus (Figure 4.7.1). At least 100 to 200 cells should be examined per specimen for routine analysis. The number of sex chromatin bodies in a nucleus is usually one less than the number of X chro-

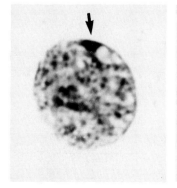 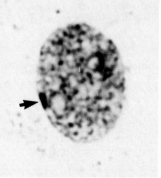

FIGURE 4.7.1. Nuclei from two cells of buccal mucosa from a female showing "Barr bodies" (*arrows*).

mosomes. Sex chromatin is generally not seen in every normal female cell but is found in a fraction of the total cells. In epithelial cells of buccal smear from normal females, for example, the incidence of sex chromatin varies between different individuals and usually ranges from 15 to 30%.

The frequency of cells exhibiting sex chromatin varies even more significantly in normal diploid female fibroblast cell cultures. The frequency of cells with sex chromatin increases with increasing cell density.

Mechanism

The sex chromatin represents the inactive X chromosome in female cells. The inactivation process in some occurs at a very early embryonic stage and remains so in all normal somatic cells. An exception is the female ovarian cell in which both X chromosomes are active and therefore have no sex chromatin. The inactive X or the so-called lyonized X remains condensed during interphase stage, which is differentially stained as the sex chromatin body.

4.72 Y-CHROMATIN

The application of quinacrine as the fluorescent stain for detection of the Y chromosome was described by Pearson et al (1970).

Protocol 4.7.2 Y-Chromatin

Solutions

1. McIlvaine's buffer (pH 5.6)
   Same as in protocol 3.21.1.
2. Quinacrine staining solution
   Same as in protocol 3.21.1.

Procedure

1. Prepare the buccal smear slides as described in protocol 4.7.1 following steps 1 through 3.
2. Place the slide in absolute methanol for 10 to 15 minutes.
3. Following fixation, let it stand in a slanting position and allow it to dry.
4. Stain the slide in quinacrine staining solution for 5 to 10 minutes.
5. Rinse in running tap water.
6. Mount in McIlvaine's buffer (pH 5.6).
7. Examine the slide using a fluorescence microscope with a wavelength of 450 to 500 nm.

Staining Pattern

The Y-chromatin body is seen as a brilliantly fluorescent body in the nucleus (Figure 4.7.2). At least 100 cells should be examined for analysis.

Mechanism

The mechanism involved in differential fluorescence by quinacrine has been described in section 3.1.1. The heterochromatic distal long arm region of the Y chromosome contains satellite DNAs and exhibits brilliant fluorescence by quinacrine staining. This heterochromatin region can be seen as a brilliant fluorescent body in interphase nuclei.

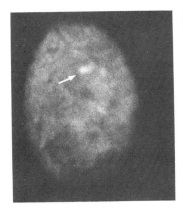 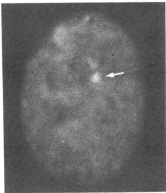

FIGURE 4.7.2. Nuclei from two cells of buccal mucosa from a male showing "Y-bodies" (arrows).

Applications

The X- and Y-chromatin studies are useful for rapid screening of clinical specimens to tentatively diagnose the possible sex chromosome constitution (Rary et al, 1978; Quinsheng et al, 1980). Sex chromatin staining has been used to investigate the nature and organization of chromatin bodies in cells containing structurally abnormal X chromosomes (Mutchnik et al, 1981). Sex chromatin studies in female breast carcinoma cell lines have shown the absence of inactive X chromosome and two active X chromosomes (Camargo and Wang, 1980). The sex chromatin studies using Barr body staining have been used more frequently than has Y-body staining. It was suggested that the frequency of Y-chromatin in males undergoes a transition during early infancy (Jijima et al, 1978; Moraitou et al, 1981; Welch et al, 1974). Caution must be exercised in Y-chromatin studies, because the fluorescence nature of the small Y and the heteromorphisms of short arms of acrocentric chromosomes may overlap and mislead the investigator.

4.8   REFERENCES

Angell RR, Jacobs PA: Lateral asymmetry in human constitutive heterochromatin. *Chromosoma* 1975; 51:301–310.

Babu A, Chemitiganti S, Verma RS: Heterochromatinization of human X chromosomes: Classification of replication profiles. *Clin Genet* 1986;30:108–111.

Barr ML, Bertram LF: A morphological distinction between neurones of the male and the female and the behavior of the nucleolar satellite during accelerated nucleoprotein synthesis. *Nature* 1949;163: 676–677.

Berger R, Bloomfield CD, Sutherland GR: Report of the committee on chromosome rearrangements in neoplasia and on fragile sites. Eighth International Workshop on Human Gene Mapping. *Cytogenet Cell Genet* 1985;40:490–535.

Brito-Babapulle V: Lateral asymmetry in human chromosomes 1, 3, 4, 15, and 16. *Cytogenet Cell Genet* 1981;29:198–202.

Camargo M, Cervenka J: Pattern of chromosomal replication in synchronized lymphocytes. 1. Evaluation and application of methotrexate block. *Hum Genet* 1980;54:47–53.

Camargo M, Wang N: Cytogenetic evidence for the absence of an inactivated X chromosome in a human female (XX) breast carcinoma cell line. *Hum Genet* 1980;55:81–85.

Carrano AV, Minkler JL, Stetka DG, Moore II DH: Variation in the baseline sister chromatid exchange frequency in human lymphocytes. *Environ Mutag* 1980;2:325–337.

Chaganti RSK, Schonberg S, German J: A many-fold increase in sister chromatid exchanges in Bloom's syndrome lymphocytes. *Proc Natl Acad Sci USA* 1974;71:4508–4512.

Cheung SW, Crane JP, Burgess AC: High resolution R banding in amniotic fluid cells using BrdU-Hoechst 33258-Giemsa (RBG) technique. *Hum Genet* 1985;69:86–87.

Comings DE: Mechanisms of chromosome banding and implications for chromosome structure. *Ann Rev Genet* 1978;12:25–46.

Craig-Holmes AP, Shaw MW: Cell cycle analysis of asynchronous cultures using the BUdR-Hoechst technique. *Exp Cell Res* 1976;99:79–87.

Croci G: BrdU sensitive fragile site on the long arm of chromosome 16. *Am J Hum Genet* 1983;35: 530–533.

Crossen PE, Morgan WF: Analysis of human lymphocyte cell cycle time in culture measured by sister chromatid differential staining. *Exp Cell Res* 1977;104:453–547.

deArce MA: The effect of donor sex and age on the number of sister chromatid exchanges in human lymphocytes growing in vitro. *Hum Genet* 1981;57:83–85.

DeBraekeleer M, Smith B, Lin CC: Fragile sites and structural rearrangements in cancer. *Hum Genet* 1984;69:112–116.

Dutrillaux B, Laurent C, Couturier J, Lejeune J: Coloration par l'acridine orange de chromosomes prealablement traités par le 5-bromodeoxyuridine (BUdR). *C R Hebd Seances Acad Sci* 1973;276: 3179–3181.

Dutrillaux B, Couturier J, Richter C-L, Viegas-Pequignot E: Sequence of DNA replication in 277 R- and Q-bands of human chromosomes using BrdU treatment. *Chromosoma* 1976;58:51–61.

Flamm WG, McCallum M, Walker PMB: The isolation of complementary strands from mouse DNA fraction. *Proc Natl Acad Sci USA* 1967;57:1729–1734.

Friedman JM, Howard-Peebles PN: Inheritance of fragile X syndrome: An hypothesis. *Am J Med Genet* 1986;23:701–714.

Gallow JH, Ordonez JV, Brown GE, Testa JR: Synchronization of human leukemic cells: Relevance for high-resolution chromosome banding. *Hum Genet* 1984;66:220–224.

Galloway SM, Evans HJ: Asymmetrical C-bands and satellite DNA in man. *Exp Cell Res* 1975;94:454–459.

Gebhart E: Sister chromatid exchange (SCE) and structural chromosome aberration in mutagenicity testing. *Hum Genet* 1981;58:235–254.

Ghosh PK, Rani R, Nand R: Lateral asymmetry of constitutive heterochromatin in human chromosomes. *Hum Genet* 1979;52:79–84.

Glover TW: FUdR induction of the X chromosome fragile site: Evidence for the mechanism of folic acid and thymidine inhibition. *Am J Hum Genet* 1981;33:234–242.

Glover TW, Berger C, Coyle J, Echo B: DNA polymerase inhibition by aphidicolin induces gaps and breaks at common fragile sites in human chromosomes. *Hum Genet* 1984;67:136–142.

Haaf T, Ott G, Schmid M: Differential inhibition of sister chromatid condensation induced by 5-azadeoxycytidine in human chromosomes. *Chromosoma* 1986;94:389–394.

Hecht F: The fragile site hypothesis of cancer (Editorial). *Cancer Genet Cytogenet* 1988a;31:119–121.

Hecht F: Fragile sites update (Appendix A). *Cancer Genet Cytogenet* 1988b;31:125–128.

Hecht F, Jacky PB, Sutherland GR: The fragile X chromosome: Current methods. *Am J Med Genet* 1982;11:489–495.

Hilwig I, Gropp A: Decondensation of constitutive heterochromatin in L cell chromosomes by a benzimidazole compound ("33258-Hoechst"). *Exp Cell Res* 1973;81:474–477.

Hoegerman SF, Rary JM: Speculation on the role of transposable elements in human genetic disease with particular attention to achondroplasia and fragile X syndrome. *Am J Med Genet* 1986;23:685–699.

Howard-Peebles PN, Pryor JC: Fragile sites in human chromosomes. I. The effect of methionine on the Xq fragile site. *Clin Genet* 1981;19:228–232.

Jacky PB, Sutherland GR: Thymidylate synthetase inhibitors and fragile-site expression in lymphocytes. *Am J Hum Genet* 1983;35:1276–1283.

Jacobs PA, Glover TW, Mayer M, Fox P, Gerrard JW, Dunn HG, Herbst DS: X-linked mental retardation: A study of 7 families. *Am J Med Genet* 1980;7:471–489.

Jenkins EC, Brown WT, Duncan CJ, Brooks J, Ben-Yishay M, Giordano FM, Nitowsky HM: Feasibility of fragile X chromosome prenatal diagnosis demonstrated (letter). *Lancet* 1981;2:1292.

Jenkins EC, Brown WT, Brooks J, Duncan CJ, Rudelli RD, Wisniewski HM: Experience with prenatal fragile X detection. *Am J Med Genet* 1984;17:215–239.

Jenkins EC, Brown WT, Wilson MG, Lin MS, Alfi OS, Wassman ER, Brooks J, Duncan CJ, Masia A, Krawczun MS: The prenatal detection of the fragile X chromosome: Review of recent experience. *Am J Med Genet* 1986;23:297–312.

Jenkins EC, Kastin BR, Krawczun MS, Lele PK, Silverman WP, Brown WT: Fragile X chromosome frequency is consistent temporally within replicate cultures. *Am J Med Genet* 1986;23:475–482.

Jijima K, Higurashi M, Hirayama M: The transition in frequency of Y-chromatin in males during the neonatal period. *Hum Genet* 1978;42:241–243.

Kaluzewski B: BrdU-Hoechst-Giemsa analysis of DNA replication in synchronized lymphocyte cultures. Study of human X and Y chromosomes. *Chromosoma* 1982;85:553–569.

Kim MA: Fluorometrical detection of thymine base differences in complementary strands of satellite DNA. *Humangenetik* 1975;28:57–63.

Krawczun MS, Lele PK, Jenkins EC, Brown WT: Fragile X expression increased by low cell-culture density. *Am J Med Genet* 1986;23:467–474.

Kucherlapati RS, Hilwig I, Gropp A, Ruddle FH: Mammalian chromosome identification in interspecific hybrid cells using "Hoechst 33258." *Humangenetik* 1975;27:9–14.

Latt SA: Microfluorometric detection of deoxyribonucleic acid replication in human metaphase chromosomes. *Proc Natl Acad Sci USA* 1973;70:3395–3399.

Latt SA: Sister chromatid exchanges, indices of human chromosome damage and repair: Detection of fluorescence and induction of mitomycin C. *Proc Natl Acad Sci USA* 1974;71:3162–3166.

Latt SA, Davidson RL, Lin MS, Gerald PS: Lateral asymmetry in the fluorescence of human Y chromosomes stained with 33258 Hoechst. *Exp Cell Res* 1974;87:425–429.

LeBeau MM: Chromosomal fragile sites and cancer-specific breakpoints — A moderating view (editorial). *Cancer Genet Cytogenet* 1988;31:55–61.

Lin MS, Alfi OS: Detection of sister chromatid exchanges by 4′-6-diamidino-2-phenylindole fluorescence. *Chromosoma* 1976;57:219–225.

Lin MS, Latt SA, Davidson RL: Microfluormetric detection of asymmetry in the centromeric region of mouse chromosomes. *Exp Cell Res* 1974;86:392–395.

Lubs HA: A marker X chromosome. *Am J Hum Genet* 1969;21:231–244.

Lyon MF: Sex chromatin and gene action in the mammalian X-chromosome. *Am J Hum Genet* 1962;14:135–148.

Marcus M, Goitein R, Gropp A: Condensation of all human chromosomes in phase G_2 and early mitosis can be drastically inhibited by 33258-Hoechst treatment. *Hum Genet* 1979a;51:99–105.

Marcus M, Nattenberg A, Goitein R, Nielsen K, Gropp A: Inhibition of condensation of human Y chromosome by the fluorochrome Hoechst 33258 in a mouse-human cell hybrid. *Hum Genet* 1979b;46:193–198.

Mattei MG, Mattei JF, Vidal I, Giraud F: Expression in lymphocyte and fibroblast culture of the fragile X chromosome: A new technical approach. *Hum Genet* 1981;59:166–169.

Moraitou EL-, Kourounis GS, Kostaraki PG-, Lyberators C, Aravantinos D, Kaskarelis D: The transition in frequency of Y-chromatin in males during early infancy. *Clin Genet* 1981;20:416–418.

Mutchinik O, Casas L, Ruz L, Lisker R, Lozano O: Symmetrical replication patterns and sex chromatin bodies formation of an idic(X) (p22.3::p22.3) chromosome. *Hum Genet* 1981;57:261–264.

Nussbaum RL, Airhart SD, Ledbetter DH: Recombination and amplification of pyrimidine-rich sequences may be responsible for initiation and progression of the Xq27 fragile site: An hypothesis. *Am J Med Genet* 1986;23:715–722.

Pai GS, Thomas GH: A new R-banding technique in clinical cytogenetics. *Hum Genet* 1980;54:41–45.

Painter RB: A replication model for sister-chromatid exchange. *Mutat Res* 1980;70:337–341.

Pearson PL, Bobrow M, Vosa CG: Technique for identifying Y chromosomes in human interphase nuclei. *Nature* 1970;226:78–79.

Pembrey ME, Winter RM, Davies KE: A premutation that generates a defect at crossing over explains the inheritance of fragile X mental retardation. *Am J Med Genet* 1985;21:709–717.

Perry P, Wolff S: New Giemsa method for differential staining of sister chromatids. *Nature* 1974;261:156–158.

Pimpinelli S, Prantera G, Rocchi A, Gatti M: Effects of Hoechst 33258 on human leukocytes in vitro. *Cytogenet Cell Genet* 1976;17:114–121.

Prantera G, Pimpinelli S, Rocchi A: Effects of distamycin A on human leukocytes in vitro. *Cytogenet Cell Genet* 1979;23:103–107.

Prantera G, DiCastro M, Cipriani L, Rocchi A: Inhibition of human chromosome condensation induced by DApi as related to cell cycle. *Exp Cell Res* 1981;101:235–243.

Quinsheng G, Liyu H, Minyi T, Yiwen Z, Houshun T, Jinxia L, Xienting Z: Sex chromosome and chromatin examination in gynecology. *Gynecol Obstet Invest* 1980;11:17–36.

Rary JM, Cummings D, Jones HW: The fallability of X-chromatin as a screening test for anomalies of the X chromosome. *Obstet Gynecol* 1978;51:107–108.

San RHC, Stich W, Stich HF: Differential sensitivity of xeroderma pigmentosum cells of different repair capacities towards the chromosome breaking action of carcinogens and mutagens. *Int J Cancer* 1977;20:181–187.

Scheres JMJC, Hustinx TWJ: Heritable fragile sites and lymphocyte culture medium containing BrdU. *Am J Hum Genet* 1980;32:628–629.

Schmid M: Demonstration of Y/autosomal translocations using distamycin A. *Hum Genet* 1979;53:107–109.

Schmid M, Klett C, Niederhofer A: Demonstration of a heritable fragile site in human chromosome 16 with distamycin A. *Cytogenet Cell Genet* 1980;28:87–94.

Schmid M, Grunert D, Haaf T, Engel W: A direct demonstration of somatically paired heterochromatin of human chromosomes. *Cytogenet Cell Genet* 1983a;36:554–561.

Schmid M, Schmidtke J, Kruse K, Tolksdorf M: Characterization of a Y/15 translocation by banding methods, distamycin A treatment of lymphocytes and DNA restriction endonuclease analysis. *Clin Genet* 1983b;24:234–239.

Schmid M, Hungerford DA, Poppen A, Engel W: The use of distamycin A in human lymphocyte cultures. *Hum Genet* 1984a;65:377–384.

Schmid M, Haaf T, Grunert D: 5-Azacytidine-induced undercondensations in human chromosomes. *Hum Genet* 1984b;67:257–263.

Schmid M, Ott G, Haaf T, Scheres JMJC: Evolutionary conservation of fragile sites induced by 5-azacytidine and 5-azadeoxycytidine in man, gorilla, and chimpanzee. *Hum Genet* 1985;71:342–350.

Schmid M, Feichtinger W, Jesberger A, Kohler J, Lange R: The fragile site (16) (q22). I. Induction by AT-specific DNA ligands and population frequency. *Hum Genet* 1986;74:67–73.

Soudek D, Partington MW, Lawson JS: The fragile X syndrome I. Familial variation in the proportion of lymphocytes with the fragile site in males. *Am J Med Genet* 1984;17:241–252.

Shapiro LR, Wilmot PL: Prenatal diagnosis of the fra(X) syndrome. *Am J Med Genet* 1986;23:325–340.

Sutherland GR: Fragile sites on human chromosomes: Demonstration of their dependence of the type of tissue culture medium. *Science* 1977;197:265–266.

Sutherland GR: Heritable fragile sites on human chromosomes I. Factors affecting expression in lymphocyte culture. *Am J Hum Genet* 1979;31:125–135.

Sutherland GR: Fragile sites and cancer breakpoints — The pessimistic view (editorial). *Cancer Genet Cytogenet* 1988;31:5–7.

Sutherland GR, Hecht F: *Fragile Sites on Human Chromosomes*. New York, Oxford University Press, 1985.

Sutherland GR, Jacky PB: Prenatal diagnosis of fragile X chromosome. *Lancet* 1982;i:100.

Sutherland GR, Mattei JF: Report of the committee on cytogenic markers. *Cytogenetic Cell Genet* 1987;46:316–338.

Sutherland GR, Simmers RN: No statistical association between common fragile sites and non-random chromosome breakpoints in cancer cells. *Cancer Genet Cytogenet* 1988;31:9–16.

Sutherland GR, Baker E, Seshadri RS: Heritable fragile sites on human chromosomes. V. A new class of fragile site requiring BrdU for expression. *Am J Hum Genet* 1980;32:542–548.

Sutherland GR, Jacky PB, Baker EG: Heritable fragile sites on human chromosomes. X. New folate sensitive fragile sites: 6p23, 9p21, 9p32, 11q23. *Am J Hum Genet* 1983a;35:432–437.

Sutherland GR, Jacky PB, Baker EG: Heritable fragile sites on human chromosomes. XI. Factors affecting expression of fragile sites at 10q25, 16q22, and 17p12. *Am J Hum Genet* 1984;36:110–122.

Sutherland GR, Baker E, Purvis-Smith S, Hockey A, Krumins E, Eichenbaum SZ: Prenatal diagnosis of the fragile X using thymidine induction. *Prenat Diagn* 1987;7:197–202.

Tommerup N, Poulsen H, Brondum-Nielsen K: 5-fluoro-2'deoxyuridine induction of the fragile site on Xq28 associated with X linked mental retardation. *J Med Genet* 1981;18:374–376.

Tommerup N, Aula P, Gustavii B, Heiber A, Holmgren G, Koskull HV, Leisti J, Mikkelsen M, Mitelman F, Nielsen KB, Steinbach P, Ruttkowski SS, Wahlstrom J, Zang K, Zankl M: Second trimester prenatal diagnosis of the fragile X. *Am J Med Genet* 1986;23:313–325.

Viegas-Pequignot E, Dutrillaux B: Detection of G-C rich heterochromatin by 5-azacytidine in mammals. *Hum Genet* 1981;57:134–137.

Viegas-Pequignot E, Dutrillaux B: Segmentation of human chromosomes induced by 5-ACR (5-azacytidine). *Hum Genet* 1976;34:247–254.

Vogel F, Kruger J, Nielsen KB, Fryns JP, Schindler D, Schinzel A, Schmidt A, Schwinger E: Recurrent mutation pressure does not explain the prevalence of the marker (X) syndrome. *Hum Genet* 1985;71:1–6.

Wang J-CC, Erbe RW: Folate metabolism in cells from fragile X syndrome patients and carriers. *Am J Med Genet* 1984;17:303–310.

Webb TP, Rodeck CH, Nicolaides KH, Gosden CM: Prenatal diagnosis of the fragile X syndrome using fetal blood and amniotic fluid. *Prenat Diagn* 1987;7:203–214.

Webber LM, Garson OM: Fluorodeoxyuridine synchronization of bone marrow cultures. *Cancer Genet Cytogenet* 1983;8:123–132.

Welch PL, Chrisitine L, Lee Y, Wellwood H: Fluorescent Y-chromatin in newborn males. *Am J Hum Genet* 1974;26:247–251.

Willard HF: Tissue-specific heterogeneity in DNA replication patterns of human X chromosomes. *Chromosoma* 1977;61:61–73.

Willard HF, Latt SA: Analysis of deoxyribonucleic acid replication in human X chromosomes by fluorescence microscopy. *Am J Hum Genet* 1976;28:213–227.

Winter RM, Pembrey ME: Analysis of linkage relationships between genetic markers around the fragile X locus with special reference to the daughters of normal transmitting males. *Hum Genet* 1986;74:93–97.

Winter RM: Population genetics implications of the premutation hypothesis for the generation of the fragile X mental retardation gene. *Hum Genet* 1987;75:269–271.

Wolff S, Rodin B, Cleaver JE: Sister chromatid exchanges induced by mutagenic carcinogens in normal and xeroderma pigmentosum cells. *Nature* 1977;265:347–349.

Yunis JJ: Chromosomes and cancer: New nomenclature and future directions. *Hum Pathol* 1981a;12:494–503.

Yunis JJ: New chromosome techniques in the study of human neoplasia. *Hum Pathol* 1981b;12:540–549.

Yunis JJ: High resolution of human chromosomes. *Science* 1976;191:1268–1270.

Yunis JJ, Sawyer JR, Ball DW: Characterization of banding patterns of metaphase-prophase G-banded chromosomes and their use in gene mapping. *Cytogenet Cell Genet* 1978;22:679–683.

Yunis JJ, Bloomfield CD, Ensrud K: All patients with acute nonlymphocytic leukemia may have chromosomal defect. *N Engl J Med* 1981;305:135–139.

Yunis JJ, Brunning RD, Howe RB, Lobell M: High-resolution chromosomes as an independent prognostic indicator in adult acute nonlymphocytic leukemia. *N Engl J Med* 1984;311:812–818.

5

In Situ Hybridization

Paul Szabo

In situ hybridization is a technique that allows detection of nucleic acid sequences, either RNA or DNA, in cytologic preparations. Historically, the method has primarily been applied for mapping specific DNA sequences onto chromosomes. This method is based on the pioneering studies of Gall and Pardue (1969) who used labeled 18 + 28S ribosomal RNA probes to detect these genes in nucleoli of *Xenopus laevis*, and Jones (1970) who localized mouse satellite DNA to the centric heterochromatic regions of mouse chromosomes. Subsequent studies by these and other investigators extended the application to the mapping of other repeated DNA sequences in the genomes of numerous species, some with polytene chromosomes as well as those with diploid chromosomes (Wimber and Steffensen, 1970; Henderson et al, 1972; Steffenson et al, 1974; Gosden et al, 1975; Szabo et al, 1977; Yu et al, 1978; Elder et al, 1980; Burk et al, 1985). Most of these studies employed nucleic acid sequences that were abundant and thus could be purified relatively easily, such as the ribosomal RNA species or tandemly repeated satellite DNA sequences. Such sequences could be converted to a radioactive probe by using nucleic acid polymerases and labeled precursor molecules to make a radioactive copy or direct labeling by chemical modification, such as iodination with Na ^{125}I. The advent or recombinant DNA technology made it possible, at least in principle, to generate probes for any gene of interest whether it was repeated or unique. Many of the genes that were localized in the early in situ hybridization studies were found in multiple, often tandem copies, making detection easier because a large signal was generated by the many labeled probe molecules concentrated in a small chromosomal region. It was not clear from these early studies whether the method could be applied successfully for the detection of unique sequences in diploid chromosomes where the signal from a single hybrid molecule would be much lower.

The properties of the in situ hybridization reaction were determined in model studies using ^{125}I-labeled ribosomal RNAs and drosophila polytene chromosomes (Szabo et al, 1977). Although the rates of hybridization differed, the general properties of the reaction were equivalent to those of solution hybridization or hybridization to membrane-bound DNA. These results suggested that one would be able to detect any sequence, even those found in only one copy in a diploid genome if the probe had a high enough specific activity to generate a signal that could be detected autoradiographically. However, quantitative studies using very high specific activity, iodinated probes, and human chromosomes (Cote et al, 1980) suggested that it would be impossible to detect the signal from a unique gene on a diploid chromosome in a reasonable exposure time consistently. Thus, to routinely detect unique sequences in diploid chromosomes, the signal at the site of hybridization would need to be amplified.

The approach that has been used most successfully to generate an amplified signal is the formation of a network of probe molecules during the annealing reaction in hybridization solutions containing dextran sulfate (Gerhard et al, 1981; Harper and Saunders, 1981). These networks are formed when randomly nicked double-stranded probes are denatured and used for hybridization. Such probes can be generated most easily by using the nick-translation reaction to label the probe. Preformed networks of probe molecules or labeled extensions on probes have also been used effectively to amplify the signal at the site of hybridization. The large sizes of probe networks generate enough radioactive signal to be detected in a reasonable autoradiographic exposure time. This method has been applied to the localization of a large number of cloned genes, arbitary sequences, and integrated viral DNA in humans and other species with diploid chromosomes; some references are provided (Malcolm et al, 1981; Zabel et al, 1983; Donlon et al, 1983; Rabin et al, 1984; Naylor et al, 1984; Bernheim et al, 1984; Yang-Feng et al, 1986; Chua et al, 1987).

More recently, in situ hybridization was also used successfully with nonradioactive probes that can be found by tagged detector molecules. Most commonly, modified precursors (eg, biotinylated dCTP) are introduced into the probe molecule and can be found at the site of hybridization using either antibodies or proteins with high affinity for the modified nucleotide, such as avidin coupled to fluorescent dyes (Rudkin and Stollar, 1977; Longer-Safer et al, 1982; Ladegent et al, 1985). The use of nonradioactive probes is an area of great interest; however, because detection of unique sequences in diploid chromosomes currently is still most easily accomplished using radioactive probes, this chapter will concentrate on the use of radioactive probes.

5.1   SLIDE PREPARATIONS

For most human gene localization studies, it is convenient to prepare chromosomes from short-term cultures of phytohemagglutinin (PHA)-stimulated whole blood. Five milliliters of whole blood from individuals chosen because their T lymphocytes proliferate well in response to PHA are generally sufficient to yield over 100 slides. Tissue culture cells (eg, fibroblasts) can also be used as a source of chromosomes, but we have found it consistently easier to obtain high quality chromosome preparations from peripheral T lymphocytes. However, any cell culture that has a good mitotic index as well as any procedure that yields well-spread chromosomes can be used to prepare slides for in situ hybridization. Kits from Gibco are available for culturing peripheral blood, and if hybridization studies are performed only occasionally, these kits are very convenient. Gibco medium IA and medium 4 are effective for human PHA-stimulated T-cell cultures. Generally, an individual will respond well enough in one of these two media to supply slides for three to four complete mapping studies.

5.11   CLEANING AND COATING OF SLIDES

1. Clean slides in a chromic acid bath (Chromomerge) for 2 to 4 hours. Rinse off the acid by washing the slides with running tap water for at least 1 hour followed by several changes of distilled H_2O. We have found that precleaned Gold Seal slides from Clay-Adams are adequate for use without additional acid cleaning. Metaphase chromosomes can be spread directly on clean slides or the slides can be precoated to lower background caused by DNA nonspecific probe binding to glass.

2. Several procedures are outlined that can be used to coat slides to reduce background from nonspecific binding. Although effective in reducing background, these pretreatments often yield metaphases that are not as well spread as those made on clean untreated slides.
    a. The procedure of Brahic and Haase (1978) calls for a 3-hour incubation in $3 \times SSC$ containing 1 x Denhart's solution (1 x Denhart's solution is 0.02% Ficoll, 0.02% polyvinylpyrrolidone, and 0.02% bovine serum albumin) at 65°C. The slides are then rinsed quickly in distilled H_2O and fixed for 20 minutes in ethanol and acetic acid (3:1), rinsed in ethanol, and air dried. The slides can then be stored dry until needed.

b. Slides can also be incubated in 10 x Denhart's solutions at 60° to 65°C for a minimum of 5 hours, rinsed profusely in distilled water, and air dried (Gerhard et al, 1981). Air drying at 50° to 60°C for 2 to 3 hours speeds up the process and hardens the coating which can often be delicate.

c. Alternatively, chromosome spreads can be made directly on clean glass slides that can subsequently be coated with a protective protein film. The coat is applied by dipping prepared slides in a protein solution consisting of the following:

Fetal calf serum-containing 1% formalin as a preservative	2.5 ml
Acetic acid	22.5 ml
Methanol	75 ml

After immersion, the slides are dried in a vertical position for a minimum of 3 to 4 hours before use or storage.

5.12 METAPHASE CHROMOSOME SPREADS FROM WHOLE BLOOD CULTURES

1. Add 0.5 ml of whole blood to 10 ml of RPMI 1640 media supplemented with 17% fetal calf serum, 1 mM glutamine, 100 U/ml penicillin, 100 μg/ml streptomycin, 16 U/ml heparin and 5 μg/ml PHA.

2. Culture in a Falcon 25-cm^2 flask in an upright position for 96 hours in a 5% CO_2/95% air environment at 37°C.

3. After a growth period of 72 to 96 hours add colcemid (0.05 μg/ml) to the culture 1 hour before harvest; normally, 0.05 ml of colcemid stock solution (1 μg/ml) is used.

4. Collect cells by centrifugation for 5 minutes at 500 × g and discard the supernatant.

5. Gently suspend the pellet by tapping the tube with the finger. Add prewarmed hypotonic solution (0.075 M KCl) dropwise, bring to 5 ml, mix gently with a pasteur pipette, and incubate at 37°C for 15 minutes. If some tissue culture cells do not swell as readily as lymphocytes, 0.056 M KCl can be used. The time in hypotonic solution can be adjusted for a particular cell type.

6. Spin down swollen cells at 1,000 × g for 5 minutes. Withdraw the potassium chloride with a pasteur pipette except for a small amount of liquid above the pellet so that the cells do not dry out. Place cells on ice.

7. Slowly add small amounts of fixative (3:1 methanol and glacial acetic acid) and resuspend cells gently by pipetting up and down. The volume of fixative is approximately 100 to 125% that of the hypotonic medium.

8. Incubate the cells on ice for 30 minutes and spin them down at 1,000 × g for 5 minutes at 4°C. Repeat the resuspension step. The cells can stay in the fixative for a few hours or longer.

9. Spin down the cells as above and wash one more time with the fixative. At this point the cells can be spread on slides or stored at −20°C. If stored, the fixative should be changed before use for slide preparation.

10. To prepare chromosome spreads, spin down the nuclei, resuspend the pellet gently in 0.5 to 1.0 ml of fixative, and drop on ice-cold, wet slides, 2 to 3 drops per slide. (Slides are placed in ice-cold distilled water before use for preparation of the metaphase spreads.)

11. Dry slides quickly and check slides under a microscope to be sure the chromosomes are properly spread and the mitotic index is adequate.

12. Slides can be used within the day or stored desiccated at −70°C for an extended period. We have successfully used slides that have been stored in this manner for more than 1 year.

5.2   PRETREATMENT

Before denaturing chromosomal DNA and hybridization, several steps are included in this procedure to minimize background over chromosomes and to ensure adherence to the slide. These steps include a ribonuclease treatment to eliminate RNA molecules remaining in the cell that are homologus to the sequence of interest and will thus hybridize with the labeled probe, resulting in a high background near cells. Acetylation of the slides also reduces background by decreasing the net positive charge of chromosomal and other proteins through acetylation of exposed amino groups (Hayashi et al, 1978).

1. If the slides had been stored at −70°C, take them out several hours before the experiment to warm up. Dehydrate them through an ethanol series of 2x 70%, 1x 95%, and then air dry. Freshly made slides should also be dehydrated.

2. Adhesion of metaphase spreads can be improved by incubation in $2 \times$ SSC at 70°C for 30 minutes. Alternatively, slides can be dried overnight at 60°C with similar results.

3. Slides are then treated with ribonuclease A at 25°C for 1 hour to remove any endogenous hybridizable RNA species. The RNase we have used is Sigma Type 111A made up as stock solution at 20 mg/ml in H_2O and diluted to 100 μg/ml in $2 \times$ SSC for use. After RNase digestion, the slides are washed in two changes of 2x SSC to remove excess RNase.

4. The slides are then transferred to 0.1 M triethanolamine HCl buffer (pH 8.0), and during vigorous agitation acetic anhydride is added to 0.25% (v/v). Vigorous mixing is necessary because the acetic anhydride is rapidly used in the reaction. The slides are then incubated for 10 minutes at 25°C followed by dehydration through an ethanol series of 2x 70%, 1x 95% and air drying. The slides are then ready for the DNA denaturation step.

5.3   PROBE PREPARATION

A variety of methods can be used for the preparation of labeled probes for in situ hybridization. Since the advent of recombinant DNA technology, most probes used for gene localization studies have been double-stranded DNAs cloned in plasmid vectors. The most convenient method for introducing label into a double-stranded probe is nick translation (Rigby et al, 1977) using labeled precursors, either ^{125}I-dCTP, ^{3}H nucleoside triphosphates, or α-^{35}S nucleoside triphosphates. The method to be outlined generally yields labeled DNA with a single-stranded length of 600 to 1,000 base pairs, a size that is ideal for the formation of the probe networks required to efficiently detect unique sequences. DNA probes can also be prepared using short random DNA primers and the Klenow fragment of *Escherichia coli* DNA polymerase (Feinberg and Vogelstein, 1983). However, this latter method yields probes with a shorter single-stranded length, limiting the formation of probe networks. These probes are best suited for detecting sequences that are tandemly repeated.

Another method for producing labeled probes, which is most effective for labeling RNA molecules that can be purified (eg, rRNA and tRNA), is direct labeling by iodination of the C-5 position of cytidine with Na [125]I. Probes with specific activities of 1 to 5×10^8 dpm/μg can be made routinely using this method (Prensky et al, 1973; Prensky, 1976). Because the signal observed at the site of hybridization of single-stranded RNA probes is proportional to the length of the hybrids, these iodinated probes are more appropriate for localizing tandemly repeated DNA sequences. Consequently, this method has primarily been used to label ribosomal RNAs for localization studies (Steffensen, 1977).

5.31   NICK TRANSLATION

The nick-translation reaction (Figure 5.3.1) is adapted from that of Rigby et al (1977). The procedure to be presented is for nick translation with ^{125}I-dCTP precursors; similar reaction conditions can be used for both ^{3}H-labeled or α-^{35}S-labeled precursors.

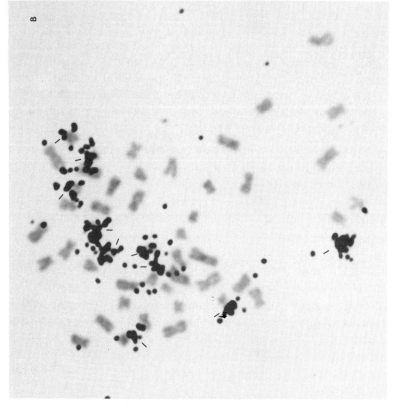

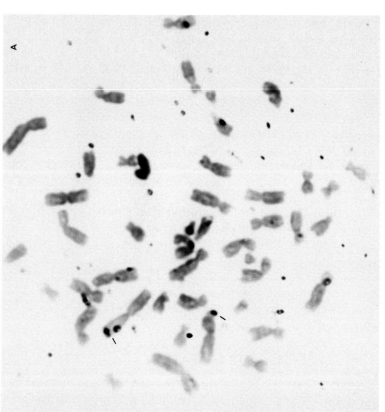

FIGURE 5.3.1. *In situ* gene localization with ^{125}I-labeled RNA probes for repeated DNA sequences. A shows the results of the hybridization of ^{125}I-labeled 5SrRNA (specific activity 10[8] dpm/μg) to human metaphase chromosomes. The 5SDNA gene site at 1q42-43 is labeled on both chromosomes 1; one of the two chromosomes has a shorter q arm because of an interstitial deletion of the interval 1q25-q32. Each 5S gene site has a target size of about 10[4] nucleotides per haploid genome. B shows the hybridization [125]I-18 +28S rRNA (specific activity 2 × 10[8] dpm/μg) to chimpanzee metaphase chromosomes. In this case, the total target size is about 1/1 × 10[5] per haploid genome spread over five gene sites. The difference in signal is a reflection of the size of the homology and differences in specific activity. Exposure time for the rRNA autoradiograph was 4 days; for the 5SrRNA experiment it was 2 weeks.

1. ^{125}I-dCTP (1,500 Ci/mM) is supplied in 50 percent ethanol and must first be concentrated by lyophilization in a microfuge tube; normally 50 to 100 μCi are used per reaction. The reaction, 30 μl of total volume, is assembled on the dried down precursor as follows:

10 x nick-translation buffer (0.5 M Tris-Cl, pH 7.5, 0.1 M MgSO$_4$, 500 mg/ml body surface area)	3 μl
0.1 M Dithiothreitol (DTT)	3 μl
Unlabeled precursors, 0.5 mM each	3 μl
200 to 400 ng of DNA	8 μl
Freshly prepared DNAase solution diluted in H$_2$O or 10 mM Tris-Cl, pH 7.5 (two serial 1/200 dilutions of the DNAase stock solution, 2 mg/ml in 50 mM Tris, pH 7.5, 10 mM MgSO$_4$ 1 mM DTT 50 μg/ml bovine serum albumin, 50% glycerol.)	3 μl
H$_2$O	9 μl
E. coli DNA polymerase (7 to 10 U).	1 μl

The reaction is then incubated at 14°C for 90 minutes.

2. The reaction is terminated by the addition of 70 μl of stop buffer (100 mM NaCl, 20 mM EDTA, 50 mM Tris-Cl, pH 8.0, 0.1% SDS) and heating at 60° to 65°C for 5 to 10 minutes.

3. Unincorporated labeled precursor is removed by "spin dialysis" through Sepharose 6B-Cl. To assemble the column, 1 ml of a 60% (v/v) suspension of the Sepharose beads in 10 mM Tris-Cl, pH 7.4, 1 mM EDTA, 0.1% SDS is added to a disposable plastic column (Isolabs Inc.) and spun at 1,000 rpm for 5 minutes. The terminated reaction, 100 μl total volume, is applied to the packed column, and the void volume, containing the labeled DNA probe, is collected by centrifugation at 1,000 rpm for 5 minutes.

4. To remove any protein components from the nick-translated probe, the solution is extracted once with an equal volume of phenol and chloroform (1:1) after the addition of 50 μg of carrier salmon sperm DNA and NaCl to 0.5 M. The phases are separated by centrifugation in a microfuge, the aqueous phase is re-extracted with an equal volume of chloroform, and the probe is precipitated by the addition of two volumes of 95 percent ethanol. The precipitate is air dried and dissolved in a small volume (50 μl) of 10 mM Tris-Cl, pH 7.4, 1 mM EDTA (TE), which is subsequently adjusted to hybridization buffer conditions (to be described).

5.32 IODINATION PROCEDURE

1. The reaction is assembled in an 0.5-ml microfuge tube behind a leaded glass shield in a chemical hood. First, 5 μl of carrier-free Na ^{125}I (about 10^{-9} mol) are acidified by the addition of 10 μl of acidification solution (0.075N HNO$_3$, 0.042 M NaAc, pH 4.7); this is followed by the addition of 10 μl of RNA probe solution (about 5 μg), and 2 μl of the catalyst, thallic nitrate 5×10^{-3} M in 1N HNO$_3$, is added and the reaction is allowed to proceed for 15 minutes at 6°C.

2. The reaction is terminated by the addition of 0.5 ml of TNE (0.05 M Tris-HCl, pH 7.4, 0.1 M NaCl, 0.001 M EDTA) containing 10^{-2} M Na$_2$SO$_3$ and reheated at 60°C for an additional 30 minutes to reduce all volatile iodine to I$^-$ and to stabilize the labeled RNA product.

3. Unreacted Na ^{125}I is then removed by passage over CF-11 cellulose (Franklin, 1966). A small column of CF-11 (about 1 ml) is washed with 0.1 NaOH, TNE, and TNE-ethanol (65:35) for large rRNA molecules or a 50:50 solution for small RNAs. The reaction mixture is also adjusted to this TNE-ethanol concentration and loaded onto the column. The column is washed profusely with the appropriate TNE-ethanol mixture to elute all unre-

acted ^{125}I. The labeled probe can then be eluted with TNE. Carrier *E. coli* tRNA, 50 to 100 μg and NaCl to 0.5 M is added and the probe is precipitated with two volumes of ethanol, air dried, and dissolved in a minimum volume of TE before preparing hybridization buffer. The probe can be stored frozen only briefly in TE; thus, it is advantageous to dilute the probe in the desired hybridization buffer.

5.4   HYBRIDIZATION

After denaturing the chromosomal DNA, the labeled probe is applied to the slides which are then incubated under conditions that promote the formation of duplex molecules. The hybridization buffer normally employed contains 50 percent formamide and moderate salt concentrations to lower the temperature at which nucleic acid annealing occurs, permitting hybridization at moderate temperatures of 37° to 45°C (McConaughy et al, 1969). Early studies used high salt buffers (eg, 5 × SSC) and elevated temperatures (60° to 68°C) to promote hybridization (Gall and Pardue, 1969; Jones, 1970), but these conditions often adversely affected the cytology, making chromosome identification difficult after hybridization. The hybridization buffer also contains Denhart's solution and unlabeled carrier DNA to limit nonspecific binding to chromosomal preparations.

The final component of the hybridization buffer is dextran sulfate which is used to increase the rate of hybridization (Wetmur, 1975). Direct measurement of the rate of in situ hybridization shows an increase of 10 to 20 times when the buffer contains 10 percent dextran sulfate (Lederman et al, 1981). This increase in hybridization rate promotes the formation of networks of probe molecules that are branched hybrids with single-stranded regions that can anneal with other probe molecules or chromosomal DNA. When these networks anneal to their homologous gene sites, they amplify the signal at the gene site, permitting detection by autoradiography because they are much larger than the individual probe molecules. This approach has also been used to enhance the signal for hybridization to filter-bound DNA (Wahl et al, 1979). In probe network formation on cytologic preparations, care must be taken to limit the concentration of probe or the time of hybridization. High background often results from the formation of excessively large, insoluble probe networks.

The in situ rate has been measured for single-stranded probes of known complexity in polytene and diploid chromosomes (Szabo et al, 1977; Cote et al, 1980). These findings indicate that cytologic hybridization is equivalent to hybridization to DNA immobilized on filters but that the rate is three to fivefold lower. The probe concentration or the time of hybridization can be altered to maximize the hybridization at the gene site while limiting the background binding. It is prudent to employ several different concentrations of probe because network formation is very variable.

We either hybridize overnight (12 to 15 hours) using hybridization solutions at probe concentrations of 50 to 200 ng/ml or we anneal for 4 to 6 hours at 100 to 600 ng/ml. These conditions have been used successfully for many cloned probes with complexities ranging from 3 to 7 kb pairs. If the complexity of the probe is much greater, a higher concentration or time of hybridization should be employed to ensure that large enough probe networks are formed and that enough of these networks are annealed to the gene site. Generally intact plasmid DNAs containing both inserts and DNA are used as probes to increase the size of the networks.

5.41   HYBRIDIZATION BUFFER

It is convenient to prepare a high concentration probe stock solution in hybridization buffer that can be diluted with additional hybridization buffer to the desired concentrations for the in situ hybridization itself. An example of the constitution of such a stock probe solution is provided below; the final volume in this example is 1 ml.

20% (w/v) dextran sulfate in deionized formamide	0.5 ml
20% SSPE	0.1 ml
50 x Denhart's solution	0.02 ml
single-stranded salmon (or herring) sperm DNA (10 mg/ml)	0.02 ml
labeled probe in TE	0.05 ml
H_2O	0.31 ml

The level of Denhart's solution, unlabeled DNA, and the amount of probe solution can be varied depending on the requirements of the experiment. When nonspecific background is high for a particular probe, the amount of Denhart's solution or unlabeled carrier DNA can be increased and the H_2O decreased accordingly. For RNA probes, unlabeled tRNA (200 to 400 μg/ml final concentration) or homopolymers (eg, poly A and poly C) can be employed as nonspecific carriers. The final concentrations of components of the hybridization buffer are 50 percent formamide, 2x SSPE, 1 x Denhart's solution (0.02% each of Ficoll, polyvinylpyrrolidone, and BSA), 200 μg/ml single-stranded salmon sperm DNA, and 10% dextran sulfate. When the diluted probe solutions are made, DNA probes can be denatured by heating at 70° to 80°C for 10 to 20 minutes. They can then be applied to the slides directly or chilled to 0°C for later use.

5.42 HYBRIDIZATION

Denatured probe solution is added to the slides, and they are allowed to incubate at 37° to 45°C for 2 to 16 hours, depending on the probe concentration. For most studies, 20 to 50 μl of probe solution is applied per slide and covered with a coverslip; 22-mm-diameter coverslips are a convenient size. Coverslips can be siliconized to prevent chromosome preparations from sticking to them. Some procedures require sealing the edges of the coverslips with rubber cement to prevent drying during incubation. We generally do not seal coverslips during hybridization but incubate the slides in humidified chambers. These chambers are square petri dishes that can hold up to three slides on glass supports. On the bottom of the dish are sheets of absorbant paper soaked in 50% formamide in 2 x SSC. After addition of the slides, these chambers are placed in a dry incubator at 37° to 45°C for the duration of the hybridization reaction; 42°C is appropriate for most studies. In the presence of 50% formamide, 42°C is near the optimal rate for hybridization in 2 x SSPE.

For single-stranded RNA probes, hybridization can be allowed to proceed until all gene sites are saturated. The length of time to achieve saturation is a function of the complexity of the probe and is approximately 5 times the $Crt_{1/2}$ which is the product of concentration and time required to achieve half saturation (Szabo et al, 1977). The $Crt_{1/2}$ has been determined for several single-stranded RNA probes of known complexity (Szabo et al, 1977; Cote et al, 1980). For probes that are not cloned but physically purified such as RNA probes, it is important not to hybridize beyond the Crt necessary for saturation of the desired gene site. This is necessary to limit the hybridization of low level contaminants in the probe preparation. Often contaminants, such as ribosomal RNAs that are homologous to repetitive DNA sequence, yield more signal than the RNA probe of interest even if they are only found as low level contaminants in the probe.

5.5 CHROMOSOMAL DNA DENATURATION

Several approaches can be used to denature chromosomal DNA including base (Gall and Pardue, 1963; Henderson et al, 1972) or acid (Steffensen et al, 1974; Szabo et al, 1978) treatments and thermal denaturation (Jones, 1970). Thermal denaturation is usually done in formamide containing buffers (Szabo et al, 1977). Formamide alters the dielectric constant of a solution and thereby lowers the melting point of DNA approximately 0.7°C per percent of formamide (McConaughy et al, 1969). This decrease in the temperature required to melt the chromosomal DNA helps to maintain good chromosome cytology. These methods also differ in the efficiency of denaturation and affect the size of the target that can be detected. Extreme treatments and high temperatures often yield chromosomes that stain poorly, are difficult to identify, and occasionally are difficult to see, but such denaturation treatments are generally highly efficient.

The choice of method depends on the nature of the experiment; highly efficient denaturation methods are needed for unique sequence gene localization in human chromosomes. Consequently, we normally employ denaturation in 70 percent formamide: 2 x SSC at 70°C for such studies. If the DNA sequences of interest are repeated (eg, ribosomal genes, satellite DNAs), a less efficient method, such as denaturation with 0.2 N HCl, can be used to generate autoradiographs with excellent cytology. After denaturation by any of the methods, slides should be used for hybridization as soon as possible, within no more than 12 hours, because the efficiency of hybridization decreases rapidly as a function of storage time after denaturation.

Base Denaturation: Slides are incubated in 0.07 N NaOH at room temperature for 2 to 3 minutes to denature the chromosomal DNA. The denaturation reaction is terminated by dehydration through an ethanol series 3x 70%, 1x 95%, and air drying. The slides can then be hybridized.

Acid Denaturation: Slides are treated with 0.2 N HCl for 15 minutes at 25°C. They are then rinsed in 2 x SSC followed by dehydration through an ethanol series 2x 70%, 1x 95%, and air drying before hybridization.

Formamide Denaturation: Several different conditions for denaturing in formamide-containing solutions have been described. We routinely use denaturation in 70% formamide and 2 x SSC at 70°C for 2 minutes. The denaturation reaction is then quenched by dehydration, as just described. An equally efficient high temperature, formamide denaturation reaction is incubation in 60% formamide, 0.2 mM EDTA, pH 8.0, at 55°C for 4 minutes followed by dehydration.

5.6 POSTHYBRIDIZATION WASHING

After hybridization, excess unhybridized probe must be removed by extensive washing of the slides. The fidelity of the hybrids is assured by washing the slides under conditions that permit retention only of actual hybrids. The stability of hybrids in situ is equivalent to that observed for hybrids formed with DNA immobilized on filter or in solution (Szabo et al, 1977). The stringency of the washing can be altered to suit a particular experiment; for example, probes from heterologous organisms for conserved sequences can be washed at lower stringency. Alternatively, the stringency can be increased in experiments to determine the location of the member(s) of a multigene family that is (are) most homologous to a given probe. Many different washing protocols have been reported, most of which are adequate. The one to be outlined is intended for double-stranded DNA probes, is relatively simple, and yields good results. For RNA probes, the addition of a ribonuclease treatment, 100 μg/ml RNAase A at 25°C for 60 minutes, along with this washing procedure ensures hybrid specificity.

1. Coverslips are allowed to float off in 2 x SSC; we normally place the slide vertically in a container filled with 2 x SSC. If the coverslips had been sealed with rubber cement during hybridization, then a wash in xylene is required to dissolve the rubber cement before the coverslips can float off. The slides are then washed with three to five changes of 2 x SSC at 25°C for 5 to 10 minutes to remove the bulk of the unhybridized, nonspecifically bound probe molecules.

2. Nonspecifically bound probes and short hybrids are then removed by two 15-minute incubations in 2 x SSC at 65° to 70°C. This condition is about 15° to 20°C below the melt-

ing point of 50% GC content DNA and assures fidelity of the hybrid. Each 1 percent mismatching reduces the melting point of duplex DNA by approximately 1°C. The temperature of these washes can be raised or lowered depending on the needs of the experiment and the thermal stability of the specific hybrid in question.

3. The slides are then washed extensively in 5 to 10 changes of 0.1 x SSC at 25°C for 10 to 15 minutes each to remove any trace levels of nonspecifically bound probe. Normally we include a higher temperature wash in 0.1 x SSC to set the final stringency of hybrids. A 55°C wash is adequate to ensure that the hybrids that will be detected are specific for the gene of interest or for closely related members of the same gene family. After the final wash, the slides are dehydrated through an ethanol series and are ready for coating with autoradiographic emulsion for hybrid detection.

5.7 HYBRID DETECTION

Autoradiography using liquid photographic emulsion to coat slides by dipping has been the method of choice to detect in situ hybrids formed using radioactive probes. This method is fast, sensitive, and relatively easy, requiring only a light-tight darkroom and drying chamber. Different types of emulsions are produced commercially by Eastman Kodak and Ilford which differ in sensitivity, resolution, and intrinsic background. The procedure to be outlined is specific for coating with the Kodak NTB series of emulsions. The tracking emulsion, NTB-2, is best suited for unique sequence gene localization using iodinated or tritiated probes. This particular emulsion detects approximately 30% of all ^{125}I decays and 10% of all ^{3}H decays. NTB-3 is a more sensitive tracking emulsion, but it leaves longer tracks of grains, often making it difficult to assign grains to the chromosomal region of origin. Emulsions also differ in their sensitivities to particular wavelengths of visible light. For NTB-2, a dark-red safelight (eg, Wratten series 2) can be used in the darkroom, but the safelight should be at least 3 feet from the working area and not directly pointing at it. Other emulsions require different safelights and different processing. A detailed description of autoradiography and its application has been published elsewhere (Rogers, 1973).

After autoradiography, the slides must be stained to visualize the chromosomes; originally, Giemsa stain was used for human chromosomes. Although this stain makes the chromosomes clearly visible, only some human chromosomes can be unambiguously identified by size and/or arm ratios (eg, chromosomes 1, 2, 3 and 16).

Staining procedures that generate consistently identifiable banding patterns are critical for precise gene localization. Several different procedures are currently employed to stain chromosomes after in situ hybridization and autoradiography for chromosome identification. The three procedures to be outlined include G-banding with Wright's stain (Chandler and Yunis, 1978), Q-banding (Levans et al, 1974), and posthybridization trypsin G-banding (Popescu et al, 1985).

5.71 Slide Dipping

1. Autoradiographic emulsion comes as a gel that can be handled with plastic or ceramic spoons. Melt a portion of emulsion in a disposable plastic tube at 45°C in a water bath. All handling of emulsion is done either in total darkness or in a darkroom equipped with a Wratten series 2 safelight.

2. Dilute emulsion 1:1 with either 2% glycerol or filtered tap water, mix gently, and transfer to dipping tube. We use disposable, plastic slide mailers designed for two slides. This type of dipping tube conserves on the amount of emulsion required to coat the slides.

3. Place two slides back to back, dip in the emulsion (3 quick dips), drain, separate the slides, and dry them vertically in a rack kept in a dry, light-tight chamber. In 36 hours the emul-

sion is usually dry enough to yield linear grain deposition without grain resorption which occurs when there is residual moisture in the emulsion.

4. After drying, transfer the slides to small, black plastic slide boxes containing drierite. These boxes are sealed with electrical tape and placed at 4°C for exposure. It is a good idea to expose a test slide(s) for 2 to 4 days to determine whether the signal-to-noise ratio is adequate for unique gene localization. Generally, 7- to 14-day exposures are adequate for human gene localizations using ^{125}I-labeled DNA probes with specific activities of 1–5 x 10^8 dpm/mg.

5.72   AUTORADIOGRAPHIC DEVELOPMENT

1. Chill the developer solution of 1:1 (V:V) dilution of D-19 (Kodak) in water to 18°C. Also chill down fixer and water washes. Tap water should be used because distilled water often causes wrinkling of emulsion layer.

2. Develop slides in the D-19 for 2.5 minutes. Lower temperatures coupled with longer times can be used to get smaller silver grains.

3. Dip once in water for 30 seconds.

4. Fix in regular fixer (Kodak) for 4 to 5 minutes; rapid fixer should be avoided because it often causes grain loss.

5. Wash slides 2 to 3 times in water at 5 minutes per wash, dehydrate through an ethanol series (2 x at 70%, 1 x at 96%), and air dry.

5.8   G-BANDING WITH WRIGHT STAIN

1. Stock Wright stain (MCB) is prepared as an 0.25% solution in anhydrous acetone-free methanol. The solution is filtered through Watman #1 filter paper and stored in dark bottles. The stain must be allowed to cure for 1 month at 25°C or 5 days at 37°C.

2. Working stain is made immediately before use by mixing 1 volume of Wright's stock solution with 3 volumes of 0.06 M phosphate buffer, pH 6.8.

3. Each slide is stained individually to determine the ideal staining time which is generally 12 to 15 minutes for autoradiographs. The slides are then rinsed with water and air dried.

4. If the banding pattern is light or the slides are overstained, the slides can be destained and restained using the same procedure for a more appropriate staining time. The slides are destained by incubating twice in 70% ethanol, once in 1% HCl absolute ethanol, and once in methanol for 2 minutes each followed by air drying. Often the banding pattern is improved by several cycles of staining and destaining.

5.81   Q-BANDING

1. To band metaphase chromosomes with quinacrine mustard after autoradiography we flood slides with the fluorescent dye atabrine, 50 μg/ml, in distilled water. The slides are then rinsed gently with running tap water to remove excess stain.

2. The slides are incubated in citric acid phosphate buffer, pH 5 to 6, for 5 minutes and mounted in this buffer for photography. The buffer is made by mixing 22.2 ml of 0.1 M citric acid and 27.8 ml of 0.2 M dibasic sodium phosphate.

3. Because only silver grains over the fluorescent chromosomes can be seen, it is necessary to also check the metaphase spreads in phase contrast or bright field to ensure that the cell is not in an area of high background, which could affect analysis.

5.82   TRYPSIN G-BANDING

1. The trypsin banding procedure to be outlined results in loss of silver grains from the developed emulsion. Consequently, the chromosomes must be stained with Giemsa stain and good metaphase spreads prephotographed to have a record of the location of silver grains. After the destaining and air drying, the slides can be banded by trypsin treatment through the emulsion for chromosome identification.

2. A fresh stock solution of trypsin 0.03% (w/v) is prepared in phosphate-buffered saline solution (PBS) containing 0.012% EDTA. This solution is diluted with PBS: Three parts trypsin solution and seven parts PBS for newly made slides. When slides are older, more trypsin is required and a two-part trypsin to three-part PBS solution can be used. This ratio can be altered empirically to optimize the banding pattern for any set of slides.

3. The slides are dipped in the appropriate working trypsin solution for 4 to 5 minutes. The reaction is terminated by two quick dips in distilled water and air drying.

4. The slides are then stained with Giemsa or Wright's stain for 20 to 40 minutes in 0.06 M phosphate buffer, pH 6.8, as already described and the prephotographed metaphase cells relocated to identify the labeled chromosomes.

5.9   APPENDIX: STOCK SOLUTIONS

A. *20 × SSC* (3 M NaCl, 0.3 M Na citrate), pH 7.4. Add per liter of final volume in distilled water.

NaCl	175.2 g
Na citrate	88.2 g

Adjust to pH 7.4 with concentrated HCl; autoclave.

B. *20 × SSPE* (3M NaCl, 0.2 M NaH_2PO_4, 0.02 M EDTA). Add per liter of final volume in double distilled water.

NaCl	175.3 g
NaH_2PO_4	27.6 g
disodium EDTA	7.4 g

Adjust pH 7.4 with 10 N NaOH.

C. *50 × Denhardt's Solution* (1% Ficoll, 1% polyvinylpyrolidone, 1% BSA). Dissolve 10 g Ficoll and 10 g polyvinylpyrolidone per 500 ml of distilled water, autoclave, and mix with an equal volume of 2 percent BSA (10 g/500 ml), aliquot, and store at −20°C.

D. *Formamide*: Deionize formamide by stirring at room temperature with mixed bed resin (Bio-rad) for 30 to 60 minutes, using roughly 5 g per 500 ml of formamide, filter, and store in a dark bottle at −20°C.

E. *0.1M Triethanolamine*: Add 13.3 ml of triethanolamine to 986.7 ml of double-distilled water and adjust to pH 8 with concentrated HCl.

5.10   REFERENCES

Bernheim A, Berger R, Szabo P: Localization of actin related sequences by *in situ* hybridization to R-banded human chromosomes. *Chromosoma* 1984;89:163–167.

Brahic M, Haase AJ: Detection of viral sequences of low reiteration frequency by *in situ* hybridization. *Proc Natl Acad Sci USA* 1978;75:6125–6129.

Burk RD, Szabo P, O'Brien S, Nash WG, Yu L, Smith KD: Organization on chromosomal specificity of autosomal homologs of human Y chromosome repeated DNA. *Chromosoma* 1985;92:225–233.

Chandler ME, Yunis JJ: A high resolution *in situ* hybridization technique for the direct visualization of labeled G-banded early metaphase and prophase chromosomes. *Cytogenet Cell Genet* 1978;22:352–356.

Chua SC, Szabo P, Vitek A, Grzeschik KH, John M, White PC: Cloning of cDNA encoding human steroid 11B-hydrophase (P450cll). *Proc Natl Acad Sci USA* 1987;84:7193-7197.

Cote BD, Uhlenbeck OC, Steffensen DM: Quantitation of in situ hybridization of ribosomal ribonucleic acids to human diploid cells. *Chromosoma* 1980;80:349-367.

Donlon TA, Litt M, Newcom SR, Magenis RE: Localization of the restriction fragment length polymorphism D14S1 (pAW-101) to chromosome 14q32.1 leads to 32.2 by in situ hybridization. *Am J Hum Genet* 1983;35:1097-1106.

Elder RT, Szabo P, Uhlenbeck OC: In situ hybridization of three transfer RNAs to polytene chromosomes of Drosophila melanogasten. *J Mol Biol* 1980;142:1-4.

Evans HJ, Buckland RA, Pardue ML: Location of the genes coding for 18S and 28S ribosomal RNA in the human genome. *Chromosoma* 1974;48:405-426.

Feinberg AP, Vogelstein B: A technique for radiolabeling DNA restriction endonuclease fragments to high specific activity. *Anal Biochem* 1983;132:6-13.

Franklin RN: Purification and properties of the replicative intermediate of the RNA bacteriophage R17. *Proc Natl Acad Sci USA* 1966;55:1504-1511.

Gall JG, Pardue ML: Formation and detection of RNA-DNA hybrid molecules in cytological preparations. *Proc Natl Acad Sci USA* 1969;63:378-383.

Gerhard DS, Kawasaki ES, Bancroft FC, Szabo P: Localization of a unique gene by direct hybridization in situ. *Proc Natl Acad Sci USA* 1981;78:3755-3759.

Gosden JR, Mitchell AR, Buckland RA, Clayton RP, Evans HJ: The location of four human satellite DNAs on human chromosomes. *Exp Cell Res* 1975;92:148-158.

Harper ME, Saunders GF: Localization of single copy DNA sequences on G-banded human chromosomes by in situ hybridization. *Chromosoma* 1981;83:431-439.

Hayashi S, Gilham IC, Delaney AD, Tener GM: Acetylation of chromosome squashes of Drosophila melanogaster decreases the background in autoradiographs from hybridization with [^{125}I-labeled RNA. *J Histochem Cytochem* 1978;26:677-679.

Henderson AS, Warburton D, Atwood KC: Localization of ribosomal DNA in the human chromosome complement. *Proc Natl Acad Sci USA* 1972;69:3394-3398.

Jones KW: The chromosomal and nuclear location of mouse satellite DNA in individual cells. *Nature* 1970;225:912-915.

Ladegent JE, Jansen-in-de-Wal N, van-Ommen GJ, Baas F, Devijlder JJ, Van-Duijn P, Van-der-ploeg M: Chromosomal localization of a unique gene by non-autoradiographic in situ hybridization. *Nature* 1985;317:175-177.

Lederman L, Kawasaki ES, Szabo P: The rate of nucleic acid annealing to cytological preparations is increased in the presence of dextran sulfate. *Anal Biochem* 1981;117:158-163.

Longer-Safer PR, Levine M, Ward D: Immunological method for mapping genes on Drosophila polytene chromosomes. *Proc Natl Acad Sci USA* 1982;79:4381-4385.

Malcolm S, Barton P, Murphy C, Ferguson-Smith MA, Bentley DL, Rabbitts TH: Localization of human immunoglobulin Kappa light chain variable region genes to the short arm of chromosome 2 by in situ hybridization. *Proc Natl Acad Sci USA* 1982;79:4957-4961.

Malcolm S, Barton P, Murphy C, Ferguson-Smith MA: chromosomal localization of a single copy gene by in situ hybridization-human beta globin genes on the short arm of chromosome 11. *Ann Hum Genet* 1981;45:135-141.

McConaughy BL, Laird CD, McCarthy BJ: Nucleic acid reassociation in formamide. *Biochemistry* 1969; 8:3289-3295.

Naylor SL, Zabel BU, Manser T, Gesteland R, Sakaguchi AY: Localization of human Ul small nuclear RNA genes to band p36.3 of chromosome 1 by in situ hybridization. *Somatic Cell Mol Genet* 1984; 10:307-313.

Popescu NC, Amsbough SC, Swan DC, DiPaolo JA: Induction of chromosome banding by trypsin/EDTA for gene mapping by in situ hybridization. *Cytogenet Cell Genet* 1985;39:73-74.

Prensky W: The radioiodination of RNA and DNA to high specific activities, in Prescot DM (ed): *Methods in Cell Biology*, New York, Academic Press, 1976, vol 13, pp 121-152.

Prensky W, Steffensen DM, Hughes W: The use of iodinated RNA for gene localization. *Proc Natl Acad Sci USA* 1973;70:1860-1864.

Rabin M, Uhlenbeck OC, Steffensen DM, Mangel WF: Chromosomal sites of integration of simian virus 40 DNA sequences mapped by in situ hybridization in two transformed hybrid cell line. *J Virol* 1984; 49:445-451.

Rigby PWJ, Diechmann M, Rhodes C, Berg P: Labeling deoxyribonucleic acid to high specific activity in vitro by nick translation with DNA polymerase I. *J Mol Biol* 1977;113:237-251.

Rogers G: *Techniques of Autoradiography*, ed 2. New York, Elsevier, 1973.

Rudkin GT, Stollar BD: High resolution detection of DNA:RNA hybrids in situ by indirect immunofluorescence. *Nature* 1977;265:472-474.

Steffensen DM, Duffey P, Prensky W: Localization of 5S ribosomal RNA genes on human chromosome 1. *Nature* 1974;252:741–743.

Steffensen DM: Human gene localization by RNA:DNA hybridization *in situ*, in Yunis JJ (ed): *Molecular Structures of Human Chromosomes*. New York, Academic Press, 1977, pp 59–88.

Szabo P, Elder R, Steffensen DM, Uhlenbeck OC: Quantitative *in situ* hybridization of ribosomal RNA species to polytene chromosomes of Drosophila melanogaster. *J Mol Biol* 1977;115:539–563.

Szabo P, Lee MR, Elder FB, Prensky W: Localization of 5S RNA and rRNA genes in the Norway rat. *Chromosoma* 1978;65:161–172.

Wahl GM, Stern M, Stark GR: Efficient transfer of large DNA fragments from agarose gels to diazobenzylozymethyl-paper and rapid hybridization by using dextran sulfate. *Proc Natl Acad Sci USA* 1979;76:3683–3687.

Wetmur JG: Acceleration of DNA renaturation rates. *Biopolymers* 1975;14:2517–2524.

Wimber DE, Steffensen DM: Localization of 5S RNA genes on Drosophila chromosomes by RNA-DNA hybridization. *Science* 1970;170:639–641.

Yang-Feng TL, Bruns GA, Carroll AJ, Simola KO, Francke U: Localization of the LDHA gene to 11p14 → 11p15 by in situ hybridization of an LDHA cDNA probe to two translocations with breakpoints in 11p13. *Hum Genet* 1986;74:331–334.

Yu L, Szabo P, Borun JW, Prensky W: Localization of the gene coding for histone H4 in human chromosomes. Cold Spring Harbor Symposium on Quant Biol 1978;XL110:1101–1105.

Zabel BU, Naylor SL, Sakaguchi AY, Bell GI, Shows TB: High resolution chromosomal localization of human genes for amylase, proopiomelanocortin, somatostatin, and a DNA fragment (D3S1) by in situ hybridization. *Proc Natl Acad Sci USA* 1983;80:6932–6936.

Blotting Techniques and Their Application

Peter Ten Dijke and Kees Stam

6.1 SOUTHERN BLOTTING

Southern blotting is a widely used analytic tool originally devised by Southern (Southern, 1975). Individual fragments of DNA, usually produced by restriction enzyme digestion, are separated by agarose gel electrophoresis and then transferred to an inert support, such as nitrocellulose. The immobilized DNA fragments can then be analyzed by hybridization with a radioactive labeled probe.

Described in this section are the techniques for DNA isolation, restriction enzyme digestion, gel electrophoresis, blotting, probe labeling, and hybridization.

6.11 DNA ISOLATION

Principle and General Notes

This method describes the isolation of high molecular weight DNA from peripheral blood, which is suitable for restriction enzyme digestion and Southern blot analysis. Prior to isolation of DNA from peripheral blood, white blood cells are purified. When cultured cells or tissue is chosen as the starting material, these cells should be treated exactly like purified white blood cells to obtain the high molecular weight DNA.

Equipment

- Polypropylene centrifuge tubes
- Plastic serologic pipettes (10 ml, 25 ml)
- Low speed centrifuge with rotor
- Pasteur pipettes
- Ultraviolet spectrophotometer

Reagents, Buffers, and Solutions

Solutions labeled DNase free must be autoclaved or made from autoclaved stocks.

- Histopaque-1119 (Sigma Diagnostics #119-1)
- RNase A, 10 mg/ml
- Proteinase K, 20 mg/ml
- 2.0 M Na acetate, pH 5.5 (DNase free)

- SE (0.15 M NaCl, 0.1 M EDTA, pH 8.0) (DNase free)
- Cell lysis buffer (10% SDS, 10 mM Tris-HCl, pH 7.5) (DNase free)
- Phenol:chloroform:isoamylalcohol (25:24:1)
- 10 × TNE (1.0 M NaCl, 0.5 M Tris-HCl, 0.05 M EDTA, pH 7.5) (DNase free)
- PBS (137 mM NaCl, 2.7 mM KCl, 1.5 mM KH_2PO_4, 8.0 mM Na_2HPO_4)
- Ethanol

Method

1. Collect 10 ml of peripheral blood in a heparinized blood-drawing tube. Store at 4°C.

2. Add 10 ml of 1 × phosphate-buffered saline solution (PBS) to the 10 ml of blood

3. Make four gradients, each in a 15-ml centrifuge tube (Corning #25319) as follows: Add 5 ml of Histopague-1119 (Sigma Diagnostics #119-1) to each tube and carefully layer 5 ml of the diluted blood onto each gradient.

4. Spin in a table-top centrifuge (eg, Beckman TJ6) at 1200 rpm for 30 minutes at room temperature.

5. After centrifugation you will see the following: Starting from the bottom of the tube going upwards: pellet of red blood cells, histopague layer, white blood cell layer, and serum layer on the top. Carefully remove the serum and discard.

6. Remove the white blood cells into a 50-ml centrifuge tube (Corning #25330). The tube must be made of polypropylene in order to withstand the phenol:chloroform:isoamyl-alcohol (25:24:1) solution.

7. Fill the centrifuge tube with 1 × PBS.

8. Spin in a Sorvall centrifuge HS-4 rotor at 1500 rpm at 4°C for 15 minutes.

9. Using a pipette, carefully remove the 1 × PBS and resuspend the pellet in 10 ml of SE buffer.

10. Add subsequently 2.5 ml of cell lysis buffer and 10 ml of phenol:chloroform: isoamylal-cohol (25:24:1). *Note*: Phenol is a dangerous liquid that causes severe skin burns. Always wear gloves when handling phenol and avoid breathing the fumes.

11. Mix the tube contents gently but thoroughly by inverting the tube for several minutes. The high molecular weight DNA can break down during the phenolization process if you shake the tube too vigorously or if you phenolize the sample for too long.

12. Spin in a low speed centrifuge at 6,000 rpm for 15 to 30 minutes at room temperature.

13. Carefully remove the aqueous phase and transfer to a clean 50-ml centrifuge tube. Use a pipette with a wide opening at the tip (3 mm) to prevent shearing of the high molecular weight DNA. If the aqueous phase is not clear after on phenolization, the extraction should be repeated once more before precipitation of the DNA.

14. Add 2 volumes of cold ethanol (−20°C). The DNA will precipitate out in a white fibrous form.

15. Scoop up the DNA with a DNase-free glass rod.

16. Put the DNA in 20 ml of 0.1 × TNE and incubate for 15 minutes at room temperature. It is not necessary for the DNA to be completely dissolved before proceeding with the RNase A treatment.

17. Add 40 μl of 10 mg/ml RNaseA (final concentration = 20 μg/ml) and incubate at 37°C for 15 minutes.

18. Add 2 ml of 10 × TNE, 1 ml of cell lysis buffer, and 120 μl of 20 mg/ml proteinase K (final concentration = 100 μg/ml) and incubate for 30 minutes at 37°C.

19. Add 20 ml of phenol:chloroform:isoamylalcohol and phenol extract as described before.

20. Precipitate the DNA using 0.1 volume of 2.0 M NaAc at pH 5.5 and 2 volumes of cold ethanol (−20°C).

21. Scoop up the DNA with a glass rod and rinse the DNA by dripping it in 15 ml of 70% ethanol.

22. Dissolve the DNA in 5.0 ml of TE buffer by mixing the sample slowly and gently for 10 minutes. If the DNA is completely dissolved, the sample will be viscous but it will not contain any visible clumps.

23. Assay the DNA concentration by U.V. spectroscopy. An A260 reading of 1.0 is approximately 50 μg/ml. Store the high molecular weight DNA at 4°C.

6.12   RESTRICTION ENZYME DIGESTION

Principle and General Notes

Restriction endonucleases are bacterial enzymes that recognize specific sequences within double-stranded DNA. The cleavage site specificities of about 400 restriction enzymes have been characterized. The majority recognize sequences of 4 to 6 nucleotides in length, and most of them have a dyad axis of symmetry termed "palindrome." The restriction enzyme activity is defined as being the amount of enzyme needed to cut 1 μg at all specific sites in a lambda DNA in 1 hour at 37°C. Factors that affect restriction enzyme activity are purity of DNA, buffer, and temperature. The typical assay conditions contain 10 mM Tris-HCl, pH 7.4, 10 mM Mg^{2+}, and proper salt concentration and the reaction is performed at 37°C. In some cases bovine serum albumin (BSA), dithiothreitol (DTT) and spermidine are added to provide optimal enzyme activity. For more details, see Maniatis et al, 1982.

Equipment

- Microfuge
- Microtubes (DNase free)
- Water baths
- Gel electrophoresis equipment
- Ultraviolet light source

Reagents and Buffers

- Restriction enzymes
- 10 × New Mix (100 mM Tris-HCl, pH 7.5, 80 mM $MgCl_2$, 0.1% gelatin and 10 mM DTT)
- NaCl (5 M)
- Spermidine (0.1 M)
- Electrophoresis grade agarose
- Electrophoresis buffer (90 mM Tris-HCl, pH 8.0, 90 mM H_3BO_4, 2.5 mM EDTA)
- Ethidium bromide (5 mg/ml)
- Orange G loading solution (20% Ficoll 400 with orange G)
- Na acetate (2.0 M pH 5.5)
- Ethanol
- Lambda DNA

Method

1. Add the following ingredients together in a microtube at room temperature: 30 μl of 10 $\times$ New Mix, an appropriate amount of NaCl, 9 μl of 0.1 M spermidine (optional), and distilled autoclaved water to make a final reaction volume of 300 μl (including the volume of DNA and restriction enzyme to be added). Addition of spermidine to a final concentration of 3 mM aids the digestion process. However, when using spermidine together with high molecular weight DNA, salt has to be added to prevent DNA precipitation.

2. Vortex for 2 seconds.

3. Add 10 to 15 μg of high molecular weight DNA and mix by holding the micro tube in one hand and flicking the tube with the forefinger of the other hand.

4. Spin for 1 second in a microfuge.

5. Add about 40 units of restriction enzyme and mix immediately as described in step 3. Always keep the restriction enzymes cold, store them in a $-20°C$ freezer, and place them in ice immediately on removal from the freezer.

6. Spin the reaction mix for 1 second in a microfuge.

7. Remove 20 μl of each reaction mix into a separate microfuge. This will be used to check the digestion of the sample DNA.

8. Add 0.5 μg of the lambda DNA to each 20-μl aliquot.

9. Mix and spin the 20-μl aliquots as in steps 3 and 4.

10. Incubate both the main digestion reactions and the digestion check reactions at an appropriate temperature (usually at 37°C) for at least 2 hours.

11. Add sample loading buffer to the digestion checks, incubate for 5 minutes at 65°C, and run on an 0.7 percent agarose gel for 2 to 3 hours at 120 mAmps (80 V). Continue incubating the main digests at 37°C, while the checks are electrophoresing.

12. Stain the gel in electrophoresis buffer with 50 μg/ml ethidium bromide for 15 minutes.

13. Check for complete DNA digestion by exposing the gel to an ultraviolet light source. The expected lambda restriction fragments should be visible in a background smear throughout the lane, indicating a complete digest.

14. Add to each main digestion 300 μl of phenol:chloroform: isoamylalcohol (25:24:1) and mix well for approximately 1 minute.

15. Centrifuge for 5 minutes in the microcentrifuge and remove the aqueous phase to a clean microtube.

16. Precipitate the DNA by adding 30 μl of 2.0 M NaAc, pH 5.5, and 750 μl cold ethanol to the aqueous phase.

17. Mix well and freeze the sample on dry ice (approximately 5 minutes) or at $-70°C$ Revco for 15 minutes.

18. Spin the sample for 15 minutes in a microfuge at 4°C and remove the ethanol using a Pasteur pipette with a "drawn" tip.

19. Add 1 ml of 70 percent ethanol solution.

20. Spin for 5 minutes in a microcentrifuge at 4°C and remove the 70 percent ethanol.

21. Leave the tube open and allow the DNA pellet to air dry for 15 minutes at room temperature.

22. Dissolve the DNA in 50 μl of TE buffer.

23. Store the digested DNA samples on ice. They are now ready for gel electrophoresis.

6.13 AGAROSE GEL ELECTROPHORESIS

Principle and General Notes

Agarose gel electrophoresis is a standard method used to separate DNA fragments. The location of DNA within the gel can be determined by staining with the fluorescent, intercalating dye ethidium bromide. The DNA migration depends on size, agarose concentration, conformation of the DNA (circular or linear), and applied current (Maniatis et al, 1982). The agarose gel with DNA restriction fragments can be used for blotting onto nitrocellulose.

Equipment

- Microfuge and microtubes
- Horizontal (submarine) electrophoresis apparatus
- Power supply
- Plastic gel formers with comb
- Ultraviolet light source

Reagents and Buffers

- Electrophoresis grade agarose
- Ethidium bromide (5 mg/ml)
- Electrophoresis buffer (90 mM Tris-HCl pH 8.0, 90 mM H_3BO_3, 2.5 mM EDTA)
- Orange G loading solution (20% Ficoll 400 with orange G)
- DNA size markers

Method

1. Add an appropriate amount of agarose in electrophoresis buffer and boil until the solution is completely homogenous and no solid particles of agarose remain. Agarose (0.8%) is good for separation of fragments ranging in size from 1 to 25 kb. For smaller fragments higher percentage gels (1.5%) and for large fragments lower percentage gels (0.5%) are used.

2. Cool the solution to 50°C before pouring the gel. The sample wells are made with a comb that is adjusted so that the bottom of the comb is about 0.5 mm off the gel bed.

3. Let the gel solidify for approximately 60 minutes, remove the comb, and place the gel in the electrophoresis tank filled with electrophoresis buffer.

4. Add to each 50-μl sample 20 μl of orange G loading solution.

5. Mix and carefully load each sample into one well of the gel. For size estimation of the digested DNA, DNA fragments of known size should be loaded onto the gel. Synthetic DNA markers are commercially available (eg, BRL) and cover several ranges of fragment sizes.

6. Run the gel for 2 hours at 10 mA and then overnight at 25 mA.

7. Increase the amperage to 50 mA and continue electrophoresis until the orange tracing dye has moved to the other end of the gel.

8. Stain the gel in electrophoresis buffer with ethidium bromide (50 μg/ml). Ethidium bromide is a potent carcinogen; always wear gloves when handling solutions that contain it.

9. Place the gel on the ultraviolet box and take a picture of the gel.

6.14 SOUTHERN BLOTTING

Principle and General Notes

Southern blotting is a technique to transfer and immobilize DNA from its position in an agarose gel to a nitrocellulose filter. The relative positions of the DNA restriction fragments are preserved during the transfer. Following blotting, the nitrocellulose filter is hybridized with a radioactive probe to locate the position of sequences complementary to the probe.

Equipment

- Microfuge and microtubes
- Blotting tray
- Whatman 3MM paper
- Nitrocellulose (or nylon filter)
- 80°C oven

Solutions and Buffers:

- 20 × SSC (0.3 M Na citrate, 3 M NaCl, pH 7.5)
- 2 × SSC (0.03 M Na citrate, 0.3 M NaCl)
- Denaturing solution (0.5 M NaOH, 1.5 M NaCl)
- Neutralizing solution (0.5 M Tris-HCl, pH 7.5, 3.0 M NaCl)

Method

1. After electrophoresis is completed, transfer the gel to an 8 by 8 inch glass plate and cut the gel to the size of the glass plate.

2. Transfer the gel into a photographic processing tray or equivalent, containing 500 ml of denaturing solution (made fresh).

3. Gently shake the gel on a rotating platform (eg, Tek Pro Tectator Rotator) for 15 to 30 minutes at room temperature.

4. Carefully pour off this solution and repeat this step using 500 ml of fresh denaturing solution.

5. Pour off this solution and repeat this step using 500 ml of neutralizing solution.

6. Shake the gel as in step 3 and repeat this neutralizing step one more time.

7. The gel is now ready for the Southern transfer.

8. Make a blotting tray as follows:

8.1 Place four petri dishes with lids removed in the center of a $10 \times 8 \times 2\frac{1}{4}$ inch photographic processing tray or equivalent. The petri dishes should touch each other and form a square.

8.2 Place an 8 × 8 inch glass plate on top of the petri dishes.

8.3 Cut a piece of Whatman 3MM paper $9\frac{1}{2} \times 10\frac{1}{2}$ inches, make a 1-inch slit in each corner with a pair of scissors, and place it on top of the glass plate.

8.4 Pour 20 × SSC onto the Whatman paper to wet it and into the tray to a depth of approximately 1 inch.

8.5 Smooth out any air bubbles under the Whatman paper with a glass rod or a pipette. The paper should be completely damp and hanging into the 20 × SSC.

8.6 Cover the buffer in the tray with plastic wrap so that the gel will be surrounded by a watertight seal. The plastic should cover about 1 inch of the Whatman 3MM on each side of the plate.

8.7 The blotting tray is now ready for the gel. The tray can be stored at 4°C for long periods of time, providing it is covered with plastic wrap and kept within a closed plastic bag.

9. Float a piece of pre-cut nitrocellulose paper on top of a 2 × SSC solution until the underside is completely wet and then submerge the filter for 5 minutes. Always wear gloves when handling nitrocellulose, because nitrocellulose that had been touched with greasy hands will never wet.

10. Cut two pieces of 8 × 8 inch Whatman 3MM paper (the size of the gel) and wet them as in step 2.

11. Transfer the gel from the last neutralizing step.

12. Carefully slide the gel onto the damp Whatman 3MM paper. Make sure that no air bubbles are trapped between the Whatman 3MM paper and the gel.

13. Adjust the plastic wrap so that it is underneath the gel at the sides, the top, and the bottom, but not underneath any area of the gel that contains DNA.

14. Remove the nitrocellulose filter from the 2 × SSC solution and gently lay it on top of the gel. Using gloved fingers, carefully push out any air bubbles under the filter.

15. Put the two pieces of Whatman 3MM one at a time on top of the nitrocellulose. Once again smooth out any air bubbles that may be present underneath them.

16. Place a stack of paper towels 8 × 8 inches on top of the Whatman 3MM until they are 3 to 4 inches in height.

17. Put an 8 × 8 inch glass plate on top of the paper towels and weigh it down with a 500-g weight.

   Note: The objective is to create a flow of liquid from the tray through the gel, the nitrocellulose, and the paper towels. The flow of buffer elutes the DNA fragments and from the gel and immobilizes them on the nitrocellulose paper. The plastic seal around the gel prevents short circuiting of fluid between the paper towels and the 3MM paper under the gel.

18. Allow the transfer of sample DNA to proceed at 4°C for at least 12 hours. It is convenient to do this step overnight.

19. The next morning carefully remove the paper towels and Whatman 3MM papers. Using a blue ball-point pen, mark the position of each well on the nitrocellulose filter.

20. Peel off the filter and soak it in 2 × SSC at room temperature for approximately 1 minute.

21. Dry the filter on a piece of Whatman 3MM for 1 hour at room temperature and bake it in an oven at 80°C for 2 hours.

22. Using a pair of scissors, cut the nitrocellulose so that it will fit into a hybridization box.

6.15   PROBE LABELING FOR SOUTHERN BLOTS

Principle and General Notes

Nick translation is the most common technique used for uniformly labeling hybridization probes for Southern blots. It involves the simultaneous action of two enzymes: DNA polymerase I, which has both $5' \rightarrow 3'$ exonuclease and $5' \rightarrow 3'$ polymerase activity, and DNase, which introduces nicks randomly in each strand generating free 3'-hydroxyl and free 5'-phosphate groups. The overall effect is that the initial nick is translated along the DNA molecule in a $5' \rightarrow 3'$ direction, which in the presence of radioactive nucleotides leads to a uniformly labeled population of molecules. Other methods for uniform labeling are random priming labeling and *in vitro* transcription using a phage polymerase; these methods are described in section 7.2.3.

Equipment

- Microfuge and microtubes
- Water bath (15°C)
- Geiger counter
- Liquid scintillation counter
- Sephadex G-50 column (eg, Pharmacia)

Reagents and Buffers

- 10 × nick translation buffer (500 mM Tris-HCl, pH 7.8, 50 mM MgCl$_2$, 100 mM β-mercaptoethanol). Store at −20°C.
- Nick translation dNTP stocks are made up in autoclaved, double-distilled water and stored at −20°C. dGTP and dATP stocks are 0.1 mM; dCTP and TTP stocks are 0.01 mM.
- ^{32}P-dCTP and ^{32}P-TTP (ie, Amersham PB 10205 and PB 10207, 2000 to 3000 Ci/mM).
- DNAse I stock: A concentrated solution supplied by the manufacturer should be diluted to 1 μg/μl in 50% glycerol, 50 mM NaCl, 0.5 mg/μl nuclease-free BSA (ie, BRL) and stored at −20°C.
- TE buffer (10 mM Tris-HCl, pH 7.5, 1 mM EDTA).
- TES buffer (10 mM Tris-HCl, pH 7.6, 10 mM EDTA, 0.1% SDS).

Method

1. Add on ice together:
   - DNA probe (0.2 μg in 0.5 μl)
   - 10 × nick translation buffer (1.6 μl)
   - dGTP stock (0.4 μl)
   - dATP stock (0.4 μl)
   - dCTP stock (0.2 μl)
   - TTP stock (0.2 μl)
   - distilled H$_2$O to final volume of 16 μl after the addition of all components including ^{32}P-dCTP and ^{32}P-TTP.
   - ^{32}P-dCTP (1.5 to 3.0 μl)
   - ^{32}P-dTTP (1.5 to 3.0 μl)
2. Mix, add DNA polymerase I (ie, BRL #8010SA) (3 to 4 units in 0.5 μl).
3. Mix, add 0.2 μl DNAse I (diluted stock) dilute stock 0.5 μl/500 μl TE buffer before use), and incubate 1.5 hours at 16°C.
4. Add to sample 50 μl TES buffer and 25 μl phenol:chloroform: isoamylalcohol (25:24:1).
5. Mix sample very well.
6. Centrifuge 5 minutes in eppendorf microfuge.
7. Transfer waterphase (upper) to a clean microtube and separate unincorporated label from probe, using Sephadex G-50 column equilibrated with TES buffer.

8. Collect 10 fractions of 100 μl in TES buffer.

9. Pool four hottest fractions, determined by Geiger counter, and count 1 μl of pooled fractions in a liquid scintillation counter. The expected specific activity of the labeled probe should be 1×10^9 cpm/μg with a total count equaling 1×10^8 cpm.

6.16 HYBRIDIZATION OF SOUTHERN BLOTS

Principle

The rate of hybridization between radioactive probes and filterbound nucleic acids depends on temperature, salt concentration, base mismatch, fragment length, presence of organic solvents such as formamide, and the presence of inert high molecular weight polymers such as dextran sulfate. For a review see Maniatis et al (1982).

Equipment

- Shaking waterbath with heating, ie, Precision model #50
- 65°C water bath
- Radiation monitor
- DNA blotted to nitrocellulose (ie, Schleicher and Schuell BA85 or PH79). Store dry and cool, away from chemical vapors
- Whatman 3MM paper
- Plastic box (approximately 10 × 21.5 cm) with cover or, alternatively, plastic bags, ie, Ziploc plastic self-sealing bags
- Kodak XAR-2 X-ray film, 8 × 10 inch, catalog no. 1651579
- Exposure holder 8 × 10 inch, ie, Wolf stainless steel
- Cronex lightning-plus intensifying screens, 8 × 10 inch (ie, DuPont)
- Liquid X-ray developer and fixer

Reagents, Buffers, and Solutions

A. General Stocks:

- 100 × Denhardt's: 2% Ficoll 400 (ie, Sigma No. F-4375), 2% polyvinyl pyrrolidone (Sigma PVP40), 2% BSA (fraction V, Sigma no. A-4503). First add ficoll and PVP until dissolved; add BSA and stir gently. Store in aliquots at −20°C and thaw at room temperature.
- 20 × SSC (3.0 M NaCl, 0.3 M Na citrate) (DNAse free).
- 10% Na pyrophosphate.
- 10 mg/ml poly A: Dissolve 100 mg poly A (ie Boehringer) in 10 ml of 10 mM Tris-HCl, pH 7.5, and store at −20°.
- Salmon sperm DNA: Dissolve 5.0 g salmon sperm DNA (ie, Sigma D-1626) in 500 ml 0.5 N NaOH, keep in boiling water for 20 minutes at 95°C, remove, and cool to room temperature; neutralize with 5.0 N HCl (check pH; should be neutral). Extract with phenol:chloroform:isoamylalcohol (25:24:1), remove water phase, and add 2 volumes ethanol. Incubate 30 minutes at −70°C; centrifuge 20 minutes, 6,000 rpm, at 4°C, remove ethanol, and dry pellet; dissolve in 100 ml TE (10 mM Tris-HCl, pH 7.5, 1.0 mM EDTA) and measure A260. Calculate DNA concentration. Store at −20°C.
- 1.0 M Na phosphate buffer, pH 6.5: Make 1.0 M solution of NaH_2PO_4 (basic). Make 1.0 M solution of Na_2HPO_4 (acidic). Start by adding acidic solution to basic; about 250 ml acidic and 330 ml basic give a pH of 6.5 (autoclave).

- 0.5 M EDTA (weigh desired amount and add distilled water; neutralize with 10 N NaOH to pH 7.0 to 7.5) (autoclave).

- 10% SDS (solution can be autoclaved if it is made in 10 mM Tris-HCl, pH 7.5).

- Formamide: The stock supplied by the manufacturer is assumed to be 100%. MCB formamide is suitable; the pH of the formamide should be greater than 5.0.

- Yeast RNA (ie, Boehringer No. 109223). Dissolve RNA in TE (may require heating) and extract twice with phenol:chloroform: isoamylalcohol (25:24:1). Precipitate with 2.5 volumes ethanol and 1/10 volume 2.0 M NaAc, pH 5.0. Store at −70°C for 30 minutes. Spin in centrifuge to pellet, rinse pellet with 70% ethanol twice, and dissolve in TE.

- 50 percent w/v dextran sulfate (dissolve in distilled water and store at −20°C).

B. Final Solutions:

- Prehybridization Mix I: 2.5 × SSC, 10 × Denhardt's, and 0.1% SDS (for 1.0 L). Add 20 × SSC and 100 × Denhardt's, make volume 990 ml, and add 10 ml 10% SDS. Store in aliquots at −20°C. Thaw at room temperature. If thawed at higher temperature, the solution must be mixed gently often. The solution must be incubated for 2 hours at 65°C before use.

- Prehybridization Mix II/Hybridization Mix: 2.5 × SSC, 10 × Denhardt's, 0.1% SDS, 10% dextran sulfate, 0.1% Na pyrophosphate, 10 µg/ml poly A, and 50 µg/ml salmon sperm DNA (for 1.0 liter). Dissolve 100 g of dextran sulfate in 400 ml of distilled water. Add the 20 × SSC, 10% Na pyrophosphate, and 10 mg/ml poly A. Make a volume of 890 ml; add the 100 × Denhardt's while mixing gently (the solution will be very viscous). Store in aliquots at −20°C. Thaw as for Prehybridization Mix I. The solution must be kept for 2 hours at 65°C before use. The salmon sperm DNA is added just before use (prehybridization) or with the probe (in the hybridization).

- Deluxe Wash Mix I: 2.5 × SSC, 10 × Denhardt's, 0.1% SDS, 0.1% sodium pyrophosphate, 50 µg/ml salmon sperm DNA, and 10 µg/ml poly A (for 1.0 liters). Add the 20 × SSC, 10% Na pyrophosphate, salmon sperm DNA, and 10 mg/ml poly A. Make a volume of 890 ml; add 100 × Denhardt's while mixing gently and 10 ml 10% SDS. Store at −20°C. Thaw as indicated for Prehybridization Mix I. The solution should be preheated before use, either 2 hours at 65°C or more conveniently when an overnight hybridization is done, overnight at 65°C.

- Wash Mix I: 2.5 × SSC, 0.1% SDS, and 0.1% Na pyrophosphate. Store at room temperature. Preheat the solution as indicated for Deluxe Wash Mix I.

- Higher Stringency Wash Mixes:
  - 0.3 × SSC, 0.1% SDS, and 0.1% Na pyrophosphate.
  - 0.1 × SSC and 0.1% Na pyrophosphate.
  - 0.03 × SSC, 0.1% SDS, and 0.1% Na pyrophosphate. Store at room temperature and preheat as indicated for Deluxe Wash Mix.

Method

1. Float the nitrocellulose on 2 × SSC and then submerge.

2. Prehybridize the filter(s) for 2 hours in 100 ml of 65°C Prehybridization Mix I at 65°C in a gently shaking water bath in a plastic box.

3. Incubate the filter(s) for 3 hours in 100 ml of Prehybridization Mix II/Hybridization Mix at 65°C.

4. Punch three holes in the cap of the Eppendorf tube containing the nick-translated probe with an injection needle. Heat the probe together with salmon sperm DNA for 5 minutes in a boiling water bath.

5. Add 40 to 50 ml of Hybridization Mix to a plastic box and add the probe. Mix gently but well.

6. Add the prehybridized filters one by one, each time tipping the box so the filter will be covered by Hybridization Mix, take care not to include air bubbles under the filters.

7. Seal the box and hybridize at 65°C overnight in a gently shaking water bath.

8. The next day, remove the box from the water bath, remove the tape, and transfer the filters individually to 100 ml of Deluxe Wash Mix I.

9. Incubate, shaking for 30 minutes at 65°C.

10. Transfer the filters individually to 100 to 150 ml of Wash Mix I.

11. Incubate, shaking, for 10 to 20 minutes at 65°C.

12. Continue transferring the filters to fresh Wash Mix I and incubating at 65°C. This procedure is intended to remove all nonhybridized probes; volumes and wash times may be adapted to the user's preference. When no radioactivity is coming off, as judged by a hand-held monitor, the filter can either be removed, rinsed briefly in 2 × SSC (room temperature) and dried, or washed down to a higher stringency.

13. Higher stringency washings are preferred if the DNA probe used and the DNA on the blot have 100% homology, that is, if the probe is human and the DNA on the blot is total human genomic DNA. If the homology between a human DNA probe and a more distantly related DNA (ie, *Drosophila* DNA) is being examined, it is advisable to keep the blot at 2.5 × SSC, 0.1% SDS, 0.1% Na pyrophosphate and make an autoradiogram. Blots can always be washed down further after the exposure: wet the blot in 2 × SSC, transfer to the higher stringency wash mix, and incubate. The usual higher stringency washing is in 100 ml of 0.3 × SSC, 0.1% SDS, 0.1% Na pyrophosphate for 30 minutes at 65°C in a shaking water bath. Filters can be washed in the same way in steps via 0.1 × SSC, 0.1% Na pyrophosphate, 0.1% SDS, to 0.01 × SSC, 0.1% SDS, 0.1% Na pyrophosphate. For most purposes, the wash to 0.3 × SSC, 0.1% SDS, 0.1% Na pyrophosphate is sufficient.

14. Dry the blot completely on Whatman 3MM paper.

15. Mount the blot on 3MM paper.

16. Wrap in Glad Wrap and tape.

17. Expose to XAR film overnight in an X-ray holder with intensifying screen.

Note: A simple and rapid technique for surveying large numbers of DNA samples and also RNA is dot blot hybridization (Kafatos et al, 1979). Samples can either be pipetted directly onto nitrocellulose or using a vacuum filter device. The major disadvantage of dot blot hybridization compared with Southern transfer procedure is that total DNA is analyzed rather than DNA fragments generated by restriction endonuclease digestion. Therefore, no distinction can be made between actual signal and cross-hybridization of the probe to other nucleic acids.

6.2 NORTHERN BLOTTING

Northern blotting is used to give an estimate of the length of an RNA transcript and abundance in mixed RNA populations. Many of the procedures are similar to those described for DNA. One major difference is the degradation sensitivity of RNA compared with that of DNA. The RNA is separated by electrophoresis on an agarose gel under denaturing conditions and transferred to a nitrocellulose filter. Specific RNA species are detected by hybridization with a radioactive labeled probe.

6.21   RNA ISOLATION

Principle and General Notes

Total cellular RNA in a human cell consist of 80 to 85% ribosomal, 10 to 15% low molecular weight RNA (transfer RNAs, small nuclear RNAs, etc), and 1 to 5% of mRNA. mRNA is heterogeneous in both size and sequence, and nearly all carry a poly A tract at their 3' ends. Important for RNA isolation is to minimize ribonuclease activity during purification. The endogenous RNases must be inactivated quickly during lysis of the cells. This can be accomplished with protein denaturants such as phenol, chloroform, SDS, or chaotropic salts. Tissues should be homogenized immediately after removal or alternatively be quickly frozen in liquid nitrogen and stored at −70°C until used. Preferably, the use of sterile disposable plastic-ware is used above glassware. If glassware is used, it should be treated by baking at 250°C for 4 or more hours. Whenever possible, the solutions should be treated with 0.1% diethyl pyrocarbonate for at least 12 hours and autoclaved. Note that diethyl pyrocarbonate cannot be used to treat solutions containing Tris. Gloves should be worn throughout RNA isolation to prevent RNase contamination from the skin. Isolation of total RNA is described according to the LiCl-urea method (Auffray and Rougeon, 1980) followed by polyA$^+$ RNA purification using oligo dT cellulose chromatography. Other methods of isolation are described by Maniatis et al (1982).

Equipment

- Polytron homogenizer
- Centrifuge with rotor
- Microfuge and microtubes
- Pasteur pipettes and glasswool

Reagents and Buffers

- Oligo dT cellulose (eg, Pharmacia)
- Na acetate (3M, pH 5.2)
- RNA buffer (10 mM Tris-HCl, pH 7.5, 5 mM EDTA, and 0.2% SDS)
- 0.1 M NaOH, 5 mM EDTA
- Phenol:chloroform:isoamylalcohol (25:24:1)
- Ethanol
- Loading buffer I (20 mM Tris-HCl, pH 7.6, 0.5 M NaCl, 1 mM EDTA, and 0.1% SDS)
- 2 X loading buffer I
- Loading buffer II (20 mM Tris-HCl, pH 7.6, 0.1 M NaCl, 1 mM EDTA, and 0.1% SDS)
- Elution buffer (10 mM Tris-HCl, pH 7.6, 1 mM EDTA, 0.05% SDS)

Method

A. Purification of total RNA:

1. Homogenize fresh tissue samples (1 to 3 g) in 25 ml of 3M LiCl and 6 M urea on ice with a polytron homogenizer.
2. Give five bursts for 15 seconds at 30-second intervals. The tube should be kept on ice during homogenization.
3. Keep homogenate on ice for at least 3 hours.
4. Centrifuge at 10,000 g for 1 hour to pellet the RNA.

5. Discard the supernatant and add to the pellet 4 ml of a 1:1 mixture of RNA buffer (10 mM Tris-HCl, pH 7.5, 5 mM EDTA, and 0.2% SDS) and phenol:chloroform:isoamyl-alcohol (25:24:1).

6. Vortex for about 20 minutes until no clumps are visible.

7. Transfer the solution to microtubes and spin for 5 minutes in a microfuge.

8. Carefully remove the aqueous phase and transfer it to a clean microtube.

9. Add an equal volume of phenol:chloroform:isoamylalcohol (25:24:1) to the aqueous phase and vortex well.

10. Spin and transfer the aqueous phase as described in step 8.

11. Precipitate the RNA by adding 0.1 volume of 3M Na acetate (pH 5.2) and 2.5 volumes of cold ethanol.

12. Mix well and incubate the sample overnight at $-20°C$.

13. Spin the sample for 15 minutes in a microfuge at $4°C$ and remove the supernatant by careful pipetting.

14. Rinse the pellet twice with 70% ethanol.

15. Store RNA at $-70°C$ in 70% ethanol.

B. Purification of poly A^+ RNA:

1. Pour 0.5 to 1 ml of oligo dT cellulose in a Pasteur pipette with glass wool packing.

2. Wash the column with 3 volumes of sterile water and 3 volumes of 0.1 M NaOH, 5 mM EDTA.

3. Wash the column with sterile water until effluent is neutral.

4. Equilibrate the column with loading buffer I.

5. Dissolve total RNA (1 to 10 mg) in 500 μl water.

6. Heat the sample for 5 minutes at $65°C$.

7. Add 1 volume of 2 × loading buffer I to the sample and let stand at room temperature for 5 minutes.

8. Apply this 1-ml sample to the column and collect the flowthrough.

9. Heat treat at $65°C$ for 5 minutes and cool to room temperature for 10 minutes.

10. Reapply eluate to the column.

11. Wash column subsequently with 10 ml of loading buffer I and 10 ml of loading buffer II.

12. Add 5 to 10 ml of elution buffer and collect 0.4-ml fractions in microtubes.

13. Precipitate the poly A^+ RNA with 0.1 volume of Na acetate and 2.5 volumes of ethanol at $-20°C$ overnight.

14. Spin in a microfuge for 15 minutes at $4°C$ and rinse the pellet (twice) with 70% ethanol.

15. Dissolve the pellet in autoclaved H_2O and distribute it in aliquots of 5 to 10 μg per tube.

16. Precipitate and rinse as described in steps 13 and 14. Store the poly A^+ RNA in $-70%$ ethanol at $-70°C$.

6.22   RNA GEL ELECTROPHORESIS AND BLOTTING

Principle and General Notes

To measure the molecular weight of RNA and to separate RNAs of different sizes for transfer to nitrocellulose, either the RNA is denatured with glyoxal and dimethylsulfoxide before agarose gel electrophoresis or electrophoresis is performed in the presence of methyl mercuric hydroxide or formaldehyde (Maniatis et al, 1982). We describe agarose gel electrophoresis in the presence of formaldehyde. Subsequently the DNA fragments can be blotted to nitrocellulose.

Equipment

- Gel electrophoresis equipment
- Microfuge and microtubes
- Blotting tray

Reagents and Buffers

- Agarose
- RNA electrophoresis buffer
- Formaldehyde (37%)
- Formamide (deionized)
- Orange G loading solution
- 20 X SSC

Method

1. For preparation of the gel, first add 6 g of agarose (1.2%), 50 ml of RNA electrophoresis buffer, and 360 ml of water together and autoclave.
2. Let the solution cool to 60°C.
3. Add 89.4 ml of formaldehyde (37%) in the fume hood.
4. Mix and pour the gel in the fume hood.
5. Dissolve the RNA sample in 6.8 μl of H_2O and add 3 μl 10 × RNA electrophoresis buffer, 5.2 μl formaldehyde, and 15 μl deionized formamide.
6. Mix the sample and heat for 15 minutes at 55°C and put on ice.
7. Add 15 μl of orange G loading solution to the samples and carefully load them into the wells. Because denatured RNA and DNA molecules of the same size migrate with similar mobilities, it is possible to use ^{32}P-labeled DNA restriction fragments as size markers.
8. Run the gel 12 to 18 hours at 50 V until the orange G has migrated to the other end of the gel.
9. Incubate the gel for 2 × 20 minutes in 20 × SSC.
10. Blot the RNA in the agarose gel to nitrocellulose as described for Southern blotting.

6.23   PROBE LABELING FOR NORTHERN BLOTS

Principle

In a Northern blot, specific sequences homologous to a probe can be detected in a way similar to that described for Southern blots. There is essentially no difference in probe labeling for both methods. To detect low abundance in RNA species we obtained the best results with either ^{32}P-labeled RNA probes generated by transcription labeling according to the pSP riboprobe method (Melton et al, 1984) or with a random priming method (Feinberg and Vogelstein, 1983). The first method is based on the use of RNA polymerases from a number of bacteriophages (eg, *Salmonella* phage SP6) that are very specific for their own promoters in vitro. With a DNA template, runoff radioactive RNA transcripts with a very high specific activity can be generated. In the random priming method, hexanucleotides in a random sequence are hybridized to a single standard template or a heat-denatured double-stranded DNA probe. Radioactive DNA strands complementary to a template strand are produced by the large fragment of DNA polymerase extension in the presence of radioactive nucleotides.

Equipment

- Microfuge
- Water bath
- Geiger counter
- Liquid scintillation counter

Reagents and Buffers
For random priming labeling:

- ^{32}P dCTP and ^{32}P TTP
- Large fragment DNA polymerase I
- 2.5 × OLB (125 mM Tris-HCl, pH 6.6, 0.05 mM dGTP, 0.05 mM dATP, pd (NG) primers, 13.5 OD units/ml, and 1.25 mg/ml BSA
- Sephadex G-50 columns (ie, Pharmacia)

    For SP6 labeling:

- ^{32}P-GTP (ie, Amersham PB 10161, 400 Ci/mM)
- 5 × rNTP stock (stock contains 2.5 mM each of ATP, CTP, and UTP in autoclaved double-distilled water)
- 5 × RNA transcription buffer (200 mM Tris-HCl, pH 7.5, 30 mM MgCl$_2$, 10 mM spermidine, and 100 mM NaCl)
- 0.5 M DTT
- TES buffer (10 mM Tris-HCl, pH 7.6, 10 mM EDTA, 0.1% SDS)
- Phenol:chloroform:isoamylalcohol (25:24:1)

Method
Random priming labeling:

1. Add to 100 ng of DNA probe 5 μl of distilled water.
2. Boil for 5 minutes and then incubate at 37°C for 5 minutes.
3. Add to the tube 8 μl 2.5 × OLB, 2.5 μl ^{32}PdCTP and 2.5 μl, 32p TTP.
4. Incubate for 2 to 5 hours of room temperature.
5. Add 50 μl TES buffer.
6. Run aqueous phase over Sephadex G-50 column equilibrated with TES buffer.

SP6 labeling:

1. Add to an Eppendorf tube at room temperature in the following order:
    - DNA probe (0.5 μg in 1.0 μl)
    - 5 × RNA transcription buffer (2.0 μl)
    - 0.5 M DTT (0.1 μl)
    - RNase inhibitor (10 U in 0.4 μl) (ie, Promega Biotec).

2. Mix and add SP6 RNA polymerase (1.5 U in 0.1 μl).

3. Incubate for 30 minutes at 37°C.

4. Add SP6 RNA polymerase (ie, 1.5 U in 0.1 μl).

5. Mix and incubate for 30 minutes at 37°C.

6. Mix and add 0.4 μl of RNase inhibitor.

7. Mix and add 0.2 μl of RNase-free DNase (stock diluted 10-fold in TE buffer before use).

8. Mix and incubate for 10 minutes at 37°C.

9. Add 50 μl of TES buffer and 20 μl of phenol:chloroform:isoamylalcohol (25:24:1).

10. Mix well and centrifuge for 5 minutes in an Eppendorf microfuge.

11. Transfer aqueous phase to a clean microfuge tube.

12. Separate the unincorporated label from ^{32}P-labeled RNA using Sephadex G-50 chromatography.

6.24   NORTHERN HYBRIDIZATION

Principle

There is essentially no difference in procedure for hybridizing Northern or Southern blots. The hybridization procedure at 55°C with formamide is the method of choice for RNA probes with RNA blots and will be described.

Equipment

- Shaking water bath with heating (ie, Precision model #50)
- 65°C water bath
- Radiation monitor
- DNA or RNA blotted to nitrocellulose (ie, Schleicher and Schuell BA85 or PH79). Store dry and cool, away from chemical vapors
- Whatman 3MM paper
- Plastic box (approximately 10 × 21.5 cm) with cover or alternatively plastic bags (ie, Ziploc plastic self-sealing bags)
- Kodak XAR-2 X-ray film, 8 × 10 inches (catalog no. 1651579)
- Exposure holder 8 × 10 inch (ie, Wolf stainless steel)
- Cronex lightning-plus intensifying screens, 8 × 10 inches (ie, DuPont)
- Liquid X-ray developer and fixer

Solutions

The general stock solutions are described in the section on Southern hybridization.
   Final solutions for 55°C hybridization:

- Prehybridization Mix: 50% formamide, 50 mM Na phosphate buffer, pH 6.5, 5 × SSC, 1 mM EDTA, 0.1% SDS, 2.5 × Denhardt's, 250 μg/ml salmon sperm DNA, 250 μg/ml yeast RNA, 10 μg/ml poly A. Add the formamide, Na phosphate buffer, SSC, EDTA, salmon sperm DNA, and yeast RNA. Increase volume up to final without the Denhardt's solution and SDS; add the Denhardt's, mix, and add the SDS. Make the solution fresh before each hybridization. Preheat 2 hours at 65°C before use.
- Hybridization Mix: Add four parts Prehybridization Mix to one part 50% w/v dextran sulfate. Preheat 2 hours at 65°C before use.

- Wash Mix: 2 × SSC, 20 mM Na phosphate buffer, pH 6.5, 1 mM EDTA, and 0.1% SDS.
- Higher Stringency Wash Mix: 0.2 × SSC, 20 mM Na phosphate buffer, pH 6.5, 1 mM EDTA, and 0.1% SDS.

Method

1. Pre-wet the nitrocellulose with 2 × SSC.
2. Prehybridize the filter(s) for 4 hours in 100 ml of 55°C Prehybridization Mix I at 55°C in a gently shaking water bath in a plastic box.
3. Add 40 to 50 ml of 55°C Hybridization Mix to a plastic box. Add probe and mix well.
4. Add the prehybridized filters one by one, each time tipping the box to cover the newly added filter with a layer of fluid; take care not to include air bubbles under the filters.
5. Seal the box with tape and hybridize at 55°C overnight in a gently shaking water bath.
6. Next day, remove box from the water bath, remove tape, and transfer the filters individually to 100 ml of Wash Mix.
7. Incubate, shaking, at 55°C for 15 to 30 minutes.
8. Transfer the filters to a fresh aliquot of Wash Mix and continue incubating.
9. Repeat procedure until no radioactivity remains in the Wash Mix, as judged by a portable radiation monitor.
10. Either briefly rinse the blot in 2 × SSC (room temperature) and proceed as described for the Southern blot hybridization procedure, or wash the blots down further.
11. Higher stringency washes include:

    - 0.3 × SSC, 20 mM Na phosphate buffer, pH 6.5, 1 mM EDTA, 0.1% SDS, for 30 minutes at 55°C; and
    - 0.3 × SSC, 20 mM Na phosphate buffer, pH 6.5, 1 mM EDTA, 0.1% SDS, for 30 minutes at 65°C.

6.3   APPLICATION OF NORTHERN AND SOUTHERN BLOT ANALYSIS

Chronic myelogenous leukemia (CML) is a malignancy of the pluripotent hematopoietic stem cell. Clonal expansion of the stem cell resulting in increased proliferation of myeloid and B-lymphoid lineages is characteristic of the chronic phase of the disease. In the later stage, the blast crisis, cells lose their ability to differentiate, resulting in malignant proliferation of myeloid or lymphoid cells. The leukemic cells undergo specific chromosomal translocation (the Philadelphia [Ph] translocation), affecting the long arms of chromosomes 9 and 22 (Rowley, 1973). The resulting 22q- or Ph chromosome characteristic of this type of neoplasm is found in over 95% of all patients with CML (Rowley, 1973; Heim et al, 1985). A Ph chromosome indistinguishable by cytogenetics from that found in CML is also observed in the leukemic cells of about 17 of 25% of adults with acute lymphoblastic leukemia (ALL) (Le Beau and Rowley, 1984; Clarkson, 1985; Champlin and Golde, 1985). A lower incidence has been reported for childhood ALL (about 5 percent) (Chessells et al, 1979; Ribeiro, et al, 1987) and acute nonlymphoblastic leukemia (ANLL) (less than 1%) (Yunis et al, 1984; Abe and Sandberg, 1979).

The Ph chromosome can be used as a prognostic indicator. Patients with CML with this marker generally show increased survival compared with patients with Ph-negative CML (Ezdinli et al, 1970; Pugh et al, 1985; Travis et al, 1986). By contrast, the presence of the Ph chromosome in the leukemic cells of patients with ALL appears to correlate with decreased survival (Ribeiro et al, 1987; Secker-Walter, 1984).

The critical molecular consequence of the Ph translocation is a specific gene fusion (Figure 6.3.1). A segment of the c-*abl* proto-oncogene, located at chromosome 9 band q34, becomes joined to a segment of the *phl* gene located at chromosome 22 band q11 (de Klein et al, 1982; Heisterkamp et al, 1982, 1983a, 1985). In Ph-positive CML the translocation breakpoint almost invariably lies within a 5.8-kb "breakpoint cluster region" of *phl*, designated here as bcr-210 (Groffen et al, 1984). Detailed mapping has revealed that translocation breakpoints in c-*abl* occur within intervening sequences (introns) located in the 5′ region of the gene, almost always upstream of c-*abl* exon 2. The breakpoints within the bcr-210 regions of *phl* occur in either of three introns separating the four small coding exons (numbered 1 to 4) in this region. The Ph (bcr-210) translocation diagrammed in Figure 6.3.1 depicts a hybrid gene with a junction between c-*abl*, in the intron bounded by exons 1B and 1A, and *phl*, in the intron bounded by exons 3 and 4 of the bcr-210 cluster region. Transcription of such a fused *phl*/c-*abl* gene, followed by RNA splicing, gives rise to an 8.5-kb polyadenylated RNA, with *phl* sequences at the 5′ end and *abl* sequences (almost invariably beginning with exon 2) at the 3′ end (Stam et al, 1985; Grosveld et al, 1986; Shtivelman et al, 1986). This mRNA encodes a fusion protein of approximately 210 kd, designated P210 *phl*/*abl* (Kloetzer et al, 1985; Konopka et al, 1984, 1985; Stam et al, 1987). Compared to the normal c-*abl* gene product, a polypeptide of 145 kd, the hybrid P210 *phl*/*abl* protein exhibits elevated, constitutive tyrosine protein kinase activity.

Molecular analysis of Ph-positive ALL reveals two classes of rearrangements, both involving *phl* and c-*abl*. One group exhibits Ph (bcr-210) translocations indistinguishable from those found in CML, as shown by chromosomal breakpoints within bcr-210, and synthesis of 8.5-kb mRNAs and P210 *phl*/*abl* fusion proteins. The second group, here designated Ph (bcr-190), has translocation breakpoints lying nearer to the 5′ end of the *phl* gene, within the large intron located between exons p1 and p2 (Hermans et al, 1987) (Figure 6.3.1). The Ph (bcr-190) translocations are associated with expression of a 7-kb chimeric *phl*/*abl* mRNA (Hermans et al, 1987), encoding a 190-kd *phl*/*abl* fusion protein (P190) with elevated tyro-

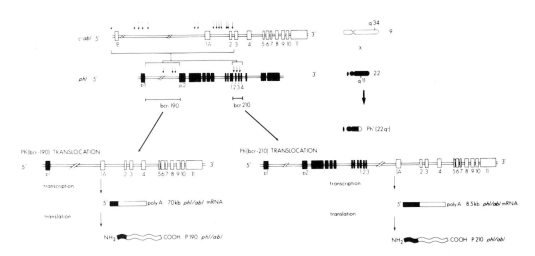

FIGURE 6.3.1. Molecular basis of the Ph translocations. Generation of Ph (22q⁻) chromosomes by recombination between *phl* and c-*abl*, and structures of spliced mRNA and of protein products. Chromosomal localization of genes and translocation events are depicted at the *right* and schematic gene structures at the *left*. Arrows indicate known positions of translocation breakpoints (individual cases for c-*abl*, representative examples for *phl*). Alternative regions in *phl* gene introns in which translocation breakpoints are found are labeled bcr-190 and bcr-210. Exons are indicated by *boxes*, introns by *double lines*. 5′ and 3′ refer to chemical polarity of mRNA or sense strand of DNA. *Wavy boxes* are the protein products (amino-terminal and carboxyl-terminal ends indicated), *open symbols* the c-*abl* sequences, and *closed symbols* the *phl* sequences.

sine protein kinase activity (Clark et al, 1987; Kurzrock et al, 1987; Chan et al, 1987; Walker et al, 1987). The P190 and P210 fusion proteins differ structurally only in the amount of *phl*-derived sequence at the aminoterminal end of the polypeptide.

The precise role of the *phl/abl* fusion proteins in leukemogenesis is not known. However, the consistent and specific presence of the P210 and P190 proteins in leukemic cells argues forcefully that their expression is an important step in the pathogenesis of Ph-positive CML and ALL.

Molecular dissection of the Ph translocation offers novel approaches to the differential diagnosis of human leukemia. The presence of the 8.5-kb *phl/c-abl* mRNA specific for chronic myelogenous leukemia can be demonstrated by Northern blot analysis. To detect this chimeric transcript a number of probes are required. We previously isolated a *phl* cDNA clone from a library of normal human fibroblast cDNA (Grosveld et al, 1986). Because the 5' and 3' ends of this cDNA are known (Okayama and Berg, 1983), orientation of the *phl* gene on chromosome 22 can be determined; its 5' end is pointed to the centromere and its 3' end to the telomere of chromosome 22. As a consequence of Ph translocation, the 5' part of the *phl* gene remains on chromosome 22, whereas the 3' part is translocated to chromosome 9 in the t(9;22) (Heisterkamp et al, 1985). Most Ph breakpoints occur in the intron sequences between the exons denoted as II and III or in the intron between exons III and IV (Figure 6.3.2A) (Heisterkamp et al, 1985). This implies that the part of the cDNA corresponding to

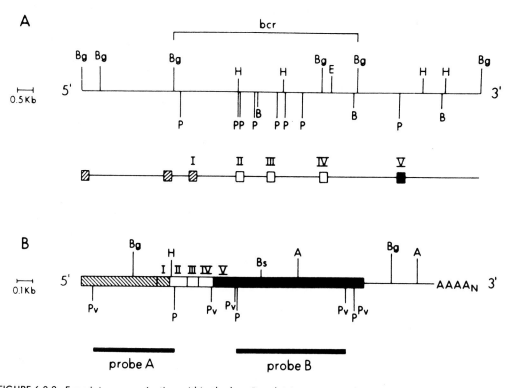

FIGURE 6.3.2. Exon-intron organization within the bcr. *Panel A* is a restriction-enzyme map of bcr. The 5.8-kb bcr, in which all breakpoints occur on chromosome 22 in the *phl* translocation, is indicated above the map. The location of exons immediately 5', 3', and within the bcr are shown (not drawn to scale) beneath the map. The numbered exons are referred to in the text. *Panel B* is a restriction-enzyme map of *phl* cDNA. The numbers and shading of regions in cDNA correspond to the exons similarly labeled in *panel A*. The *phl* cDNA probes used in this study are shown beneath the map. The following abbreviations denote the restriction enzymes used in mapping: A represents *Ava*I, B = *Bam*HI, Bg = *Bgl*II, Bs = *Bst*EII, E = *Ecor*RI, H = *Hind*III, P = *Pst*I,, and Pv = *Pvu*II. The AAA_N at the 3' end of the cDNA indicates the poly A tail.

probe A (Figure 6.3.2B) remains on chromosome 22, whereas the cDNA sequences corresponding to probe B are translocated to chromosome 9 in the Ph translocation. Both probes were inserted in reversed orientation in pSP vectors, permitting isolation of [^{32}P] RNA complementary (anti-sense) to normal *phl* mRNA. Labeled probes were generated by this vector construct, because previous experiments (Grosveld et al, 1986) had demonstrated that these probes compared with conventional nick-translated DNA probes enhance the signal at least fivefold in Northern hybridizations. Similarly, a human c-*abl* fragment encompassing the most 5′ v-*abl* homologous exon (probe C) was inserted into a pSP vector (Heisterkamp et al, 1983).

Poly A$^+$ RNA was isolated from the bone marrow cells of five patients with CML and subjected to Northern blot analysis. The hybrid transcript should hybridize with both the c-*abl* (probe C) and 5′ *phl* probe (probe A) but not with the 3′ *phl* probe (probe B), because the genomic sequences corresponding to the latter probe are translocated to chromosome 9 in t(9;22). The results of Northern blot analysis are shown in Figure 6.3.3.

All patients have an 8.5-kb transcript hybridizing with both the c-*abl* and the 5′-specific *phl* probe (Figure 6.3.3). The 3′-specific *phl* probe does not hybridize with this transcript, excluding it as a normal *phl* mRMA. Also shown are the normal 4.5- and 7.0-kb *phl* and 6.0- and 7.0-kb c-*abl* transcripts. In contrast to patients 1708, 2171, 2397, and 2252 who have the

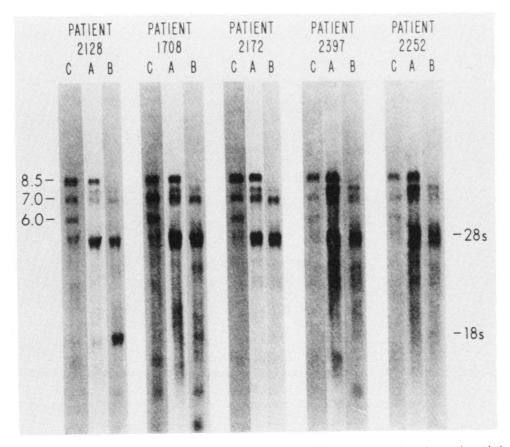

FIGURE 6.3.3. Northern blot analysis of RNA from five patients with CML. The lanes show the results with the following probes: A denotes the 5′ *phl* probe; B, the 3′ *phl* probe; and C, the c-*abl* probe. The molecular weight of c-*abl*-hybridizing RNA (in kilobases; 1 kb = 3.3 × 10^5 daltons) is indicated at the *left*; the position of ribosomal 23S and 18S RNA is shown at the *right*.

standard t(9;22) translocation, cytogenetic analysis of patient 2128 revealed a complex t(9:22),(12;22) Ph translocation. The presence of a hybrid transcript suggests that even in complex Ph translocations, *phl* and c-*abl* are fused. In a patient with an analogous karyotype (Bartram et al, 1985), both *phl* and c-*abl* sequences were found on chromosome 12 by means of in situ hybridization techniques. Northern blot analysis will likely detect the chimeric transcript in cases in which cytogenetic analysis fails to show a Ph chromosome, as in cases of a masked Ph chromosome.

The relatively small size of the bcr-210 breakpoint cluster regions enabled us to develop a DNA probe assay using Southern blot hybridization of DNA extracted from peripheral blood or bone marrow with a probe designated *phl*/bcr-3 (Blennerhassett et al, 1988) . Because of a reciprocal Ph translocation, two novel junctions should be present in genomic DNA, corresponding to the 22q$^-$ and 9q$^+$ chromosomes, respectively. If the breakpoint in *phl* lies within the bcr-210 region (Figure 6.3.1), then molecular hybridization with a DNA probe spanning this region should reveal two rearranged fragments in genomic DNA digested with an appropriate restriction endonuclease. The *phl*/bcr-3 probe (Figure 6.3.4) encompasses the entire bcr-210 region, with the exception of an internal 1.6 kb *Hind*III fragment found to contain repetitive sequences. Figure 6.3.4 indicates the human genomic DNA fragments, generated by several restriction endonuclease, that hybridize with this probe. For example, digestion with *Bgl*II normally yields three detectable fragments of 4.8, 2.3, and 1.1 kb (Fig-

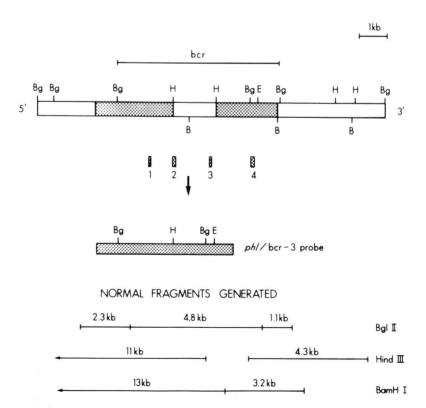

FIGURE 6.3.4. Restriction map of *phl* and construction of the *phl*/bcr-3 probe. The position of the bcr-210 breakpoint cluster region (labeled bcr) and regions of genomic DNA present in the *phl*/bcr-3 probe (*cross-hatched boxes*) are indicated. *Small boxes* labeled 1-4 indicate positions of coding exons in bcr-210 region of *phl* gene. Fragments in normal genomic DNA generated by digestion with several restriction endonucleases and detectable by the probe are indicated (size in kilobases) at the *bottom*. Restriction endonuclease cleavage sites: E = *Eco*RI, B = *Bam*HI, Bg = *Bgl*II, and H = *Hind*III.

ure 6.3.4) A translocation involving bcr-210 would disrupt one of these fragments and generate two new fragments. Because only one copy of chromosome 22 is generally rearranged in Ph'-positive leukemic cells, DNA from such cells would be expected to yield up to five DNA fragments that can hybridize with the *phl*/bcr-3 probe, that is, the three germline DNA fragments and two junction fragments.

We have found *Bgl*II to be a suitable restriction enzyme for the identification of rearranged *phl*/bcr-210 regions using the *phl*/bcr-3 probe, because of the excellent electrophoretic resolution of three hybridizable fragments generated by this enzyme and the apparent absence of polymorphism of the four relevant *Bgl*II target sites in human genomic DNA. *Bam*HI, *Xba*I, and *Eco*RI can be used as well. A rare polymorphism has been observed that affected one of the *Bam*HI sites, which may complicate the analysis using this enzyme. *Hind*III is a poor choice of enzyme for identifying Ph (bcr-210) translocations using the *phl*/bcr-3 probe, because breakpoints lying within the genomic 1.6-kb *Hind*III fragment, absent from the probe, will not be detected (Figure 6.3.4).

The results of DNA probe analysis to detect Ph (bcr-210) rearrangements in the K562 cell line and in cells of five representative patients with Ph-positive CML are shown in Figure 6.3.5. The HL-60 cell line, derived from human promyelocytic leukemia with no Ph chromosome, and bone marrow cells of a patient with Ph-negative CML served as controls lacking bcr-210 rearrangements. Genomic DNA was digested with *Bgl*II, and Southern blot hybridization with the *phl*/bcr-3 probe was carried out as described. Three germline fragments (4.8, 2.3, and 1.1 kb) are present in every case. No additional DNA fragments can be detected in DNA from either HL-60 cells (lane 1) or leukemic cells of the patient with Ph-negative

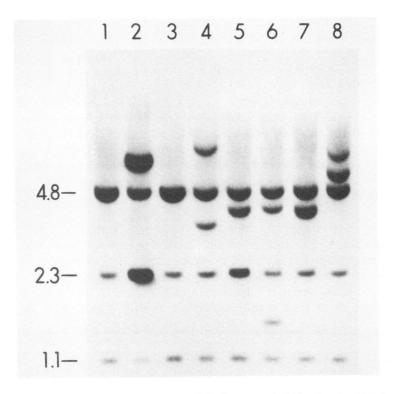

FIGURE 6.3.5. Assay for Ph (bcr-210) translocations by hybridization with *phl*/bcr-3 probe. DNA from control cell lines or patient specimens was digested with *Bgl*II and analyzed for the Ph (bcr-210) translocation by DNA probe assay as described (Methods). Sources of DNA: (1) HL-60 cell line (human acute promyelocytic leukemia); (2) K562 cell line (human CML); (3) Ph-negative CML = patient specimen; (4–8) Ph-positive CML patient specimens. Positions of germline bands (sizes in kilobases) are indicated at the *left*.

CML (lane 3). In DNA from CML cell line K562, one novel band is observed (lane 2). This band presumably represents a $22q^-$ fragment, because cytogenetic analysis revealed no $9q^+$ chromosome in these cells. The intensity of the rearranged band confirms the finding that the chimeric *phl*/c-*abl* gene is amplified in K562. In three of the samples from Ph-positive CML patients, two novel bands can be seen corresponding to the junctions present in the $22q^-$ and $9q^+$ chromosomes (lanes 4, 6, and 8). Another CML patient's sample (lane 5) also appears to have two rearranged DNA fragments, but one co-migrates with the 2.3-kb germline fragment, as judged by the intensity of the corresponding autoradiographic band. In the analysis of one of the Ph-positive leukemic DNA samples, only a single extra DNA fragment is apparent (lane 7). It is possible that a small rearranged fragment ran off the gel or that a $9q^+$ fragment is not present in these cells.

The DNA probe assay was compared with cytogenetic analysis in the diagnosis of Ph-positive leukemia in over 400 patients and in control individuals tested at seven clinical trial centers in the United States and Europe (Blennerhassett et al, 1988). In summary, the conclusion of the clinical trials is that the *phl*/bcr-210 DNA probe assay compares favorably with karyotype analysis as a specific diagnostic test for CML. The probe assay can be carried out on peripheral blood specimens as well as on bone marrow aspirates; it identifies the translocation in at least 99% of cases of cytogenetically Ph-positive CML, and it reveals a significant number of *phl*/c-*abl* translocations not detected by karyotype analysis. Furthermore, the test offers useful information in the differential diagnosis of atypical CML, CMML, and Ph-positive ALL and ANLL involving bcr-210 rearrangements, which cannot be obtained cytogenetically. As treatment of leukemia becomes increasingly sophisticated, it is highly probable that the precise diagnosis of these diseases at the molecular level will be accompanied by the development of distinct therapeutic strategies for each identifiable class.

In this chapter we discussed Southern and Northern blot techniques and their application in CML. Elucidation of the molecular structure of Ph translocation has resulted in an oncogene-based assay for the diagnosis of a specific form of human cancer. The DNA probe test is likely to find a second major use in monitoring patients with CML, particularly to determine the response to therapy. A key consideration in this application is assay sensitivity. Reconstruction experiments in which varying ratios of leukemic and nonleukemic DNA or leukemia and nonleukemic cells before DNA isolation were mixed indicate that Southern hybridization with the *phl*/bcr-3 probe test can reveal leukemic cells present at 1 percent in a peripheral blood or bone marrow cellular population. The threshold for detection of Ph-positive cells by karyotype analysis depends on the number of metaphase spreads studied per sample, but it is usually in the range of 10 percent of the cell population. Thus, under routine laboratory conditions the probe assay is several to 10-fold more sensitive than conventional cytogenetic analysis.

Recent reports demonstrate the hematologic remission of CML can be induced by recombinant human alpha-A interferon and possibly gamma interferon (Talpaz et al, 1986; Rosenblum et al, 1986; Kurzrock et al, 1987). In some patients the fraction of bone marrow cells containing the Ph chromosome decreases significantly, as demonstrated convincingly by molecular hybridization analysis (Talpaz et al, 1987; Yoffe et al, 1987). Similarly, the DNA probe assay should be useful in monitoring the response of patients with CML to other biologic response modifiers, such as the granulocyte colony-stimulating factor (Yuo et al, 1987).

Recent advances in bone marrow transplantation suggest that this approach will be increasingly important in the treatment of CML (Ganesan et al, 1987; Fefer et al, 1982; Goldman et al, 1986; McGlave et al, 1987; Punt et al, 1987). We found that *phl*/bcr-3 DNA probe assay readily revealed residual leukemic cell populations in nearly half of these patients sampled during clinical remission after transplantation. The superior sensitivity of the probe test and its precision in identifying Ph (bcr-210) translocations clearly make it a powerful tool to supplement clinical observation and cytogenetic analysis in assessing the complex biology of bone marrow transplantation.

6.4 REFERENCES

Abe S, Sandberg AA: Chromosomes and causation of human cancer and leukemia. XXXII. Unusual features of Ph1-positive acute myeloblastic leukemia, including a review of the literature. *Cancer* 1979;43:2352–2364.

Auffray C, Rougeon F: Purification of mouse immunoglobulin heavy-chain messenger RNAS from total myeloma tumor RNA. *Europ J Biochem.* 1980;107:303–314.

Bartram CR, Kleihauer E, de Klein A, Grosveld G, Teyssier JR, Heisterkamp N, Groffen J: c-abl and bcr are rearranged in a Ph1-negative CML patient. *EMBO J* 1985;4:683–686.

Ben Neriah Y, Daley GQ, Mes-Masson AM, Witte ON, Baltimore D: The chronic myelogenous leukemia-specific P210 protein is the product of the bcr/abl hybrid gene. *Science* 1986;233:212–214.

Blennerhassett GT, Furth ME, Anderson A, Burns JP, Chaganti RSK, Blick M, Talpaz M, Dev VG, Chan LC, Wiedemann LM, et al: Clinical evaluation of a DNA probe assay for the detection of the "Ph" translocation in chronic myelogenous leukemia. *Leukemia.* 1988;10:(in press).

Champlin RE, Golde DW: Chronic myelogenous leukemia: Recent advances. *Blood* 1985;65:1039–1047.

Chan LC, Karhi KK, Rayter SI, Heisterkamp N, Eridani S, Powles R, Lawler SD, Groffen J, Foulkes JG, Greaves MF, et al: A novel abl protein expressed in Philadelphia chromosome positive acute lymphoblastic leukaemia. *Nature* 1987;325:635–637.

Chessells JM, Janossy G, Lawler SD, Secker-Walker LM: The Ph1 chromosome in childhood leukaemia. *Br J Heamatol* 1979;41:25–41.

Clark SS, McLaughlin J, Crist WM, Champlin R, Witte ON: Unique forms of the abl tyrosine kinase distinguish Ph1-positive CML from Ph1-positive ALL. *Science* 1987;235:85–88.

Clarkson B: The Chronic Leukemias, in Wyngaarden JB, Smith LH (eds): Cecil Textbook of Medicine. Philadelphia, Pa, WB Saunders, 1985, pp 975–986.

de Klein A, van Kessel AG, Grosveld G, Bartram CR, Hagemeijer A, Bootsma D, Spurr NK, Heisterkamp N, Groffen J, Stepenson JR: A cellular oncogene is translocated to the Philadelphia chromosome in chronic myelocytic leukaemia. *Nature* 1982;300:765–767.

Ezdinli EZ, Sokal JE, Crosswhite L,Sandberg AA: Philadelphia-chromosome-positive and -negative chronic myelocytic leukemia. *Ann Intern Med* 1970;72:175–182.

Fefer A, Cheever MA, Greenberg MA, Greenberg PD, Appelbaum FR, Boyd CN, Buckner CD, Kaplan HG, Ranberg R, Sanders JE, Storb R, Thomas ED: Treatment of chronic granulocytic leukemia with chemoradiotherapy and transplantation of marrow from identical twins. *N Engl J Med* 1982;306: 63–68.

Feinberg AP, Vogelstein B: A technique for radiolabeling DNA restriction endonuclease fragments to high specific activity. *Anal Biochem* 1983;132:6–13.

Ganesan TS, Min GL, Goldman JM, Young BD: Molecular analysis of relapse in chronic myeloid leukemia after allogeneic bone marrow transplantation. *Blood* 1987;70:873–876.

Goldman JM, Apperley JF, Jones L, Marcus R, Goolden AW, Batchlor R, Hale G, Waldmann H, Reid CD, Hows J, et al: Bone marrow transplantation for patients with chronic myeloid leukemia. *N Engl J Med* 1986;314:202–207.

Groffen J, Stephenson JR, Heisterkamp N, de Klein A, Bartram CR, Grosveld G: Philadelphia chromosomal breakpoints are clustered within a limited region, bcr, a chromosome 22. *Cell* 1984;36: 93–99.

Grosveld G, Verwoerd T, van Agthoven T, de Klein A, Ramachandran KL, Heisterkamp N, Stam K, Groffen J: The chronic myelocytic cell line K562 contains a breakpoint in *bcr* and produces a chimeric *bcr/c-abl* transcript. *Mol Cell Biol* 1986;6:607–616.

Heim S, Billstrom R, Krostoffersson U, Mandahl N, Strombeck B, Mitelman F: Variant Ph translocations in chronic myeloid leukemia. *Cancer Genet Cytogenet* 1985;18:215–227.

Heisterkamp N, Stephenson JR, Groffen J, Hansen PF, de Klein A, Bartram CR, Grosveld G: Localization of the c-abl oncogene adjacent to a translocation break point in chronic myelocytic leukaemia. *Nature* 1983a;306:239–242.

Heisterkamp N, Groffen J, Stephenson JR, Spurr NK, Goodfellow PN, Solomon E, Carritt B, Bodmer WF: Chromosomal localization of human cellular homologues of two viral oncogenes. *Nature* 1982;299:747–749.

Heisterkamp N, Stam K, Groffen J, de Klein A, Grosveld G: Structural organization of the bcr gene and its role in the Ph′ translocation. *Nature* 1985;315:758–761.

Heisterkamp N, Groffen J, Stephenson JR: The human v-abl cellular homologue. *J Mol Appl Gen* 1983b;2:57–68.

Hermans A, Heisterkamp N, von Linden M, van Baal S, Meijer D, van der Plas D, Wiedemann LM, Groffen J, Bootsma D, Grosveld G: Unique fusion of bcr and c-abl genes in Philadelphia chromosome positive acute lymphoblastic leukemia. *Cell* 1987;51:33–40.

Kafatos FC, Jones CW, Efstra Tiadis A: Determination of nucleic acid sequence homologies and relative concentrations by a dot hybridization procedure. *Nucl Acids Res* 1979;7:1541–1552.

Kloetzer W, Kurzrock R, Smith L, Talpaz M, Spiller M, Gutterman J, Arlinghaus R: The human cellular abl gene product in the chronic myelogenous leukemia cell line K562 has an associated tyrosine protein kinase activity. *Virology* 1985;140:230–238.

Konopka JB, Watanabe SM, Witte ON: An alteration of the human c-abl protein in K562 leukemia cells unmasks associated tyrosine kinase activity. *Cell* 1984;37:1035–1042.

Konopka JB, Watanabe SM, Singer JW, Collins SJ, Witte ON: Cell lines and clinical isolates derived from Ph1-positive chronic myelongenous leukemia patients express c-abl proteins with a common structural alteration. *Proc Natl Acad Sci USA* 1985;82:1810–1814.

Kurzrock R, Shtalrid M, Romero P, Kloetzer WS, Talpaz M, Trujillo JM, Blick M, Beran M, Gutterman JU: A novel c-abl protein product in Philadelphia-positive acute lymphoblastic leukaemia. *Nature* 1987;325:631–635.

Kurzrock R, Talpaz M, Kantarfian H, Walters R, Saks S, Trujillo JM, Gutterman JU: Therapy of chronic myelogenous leukemia with recombinant interferon-gamma. *Blood* 1987;70:943–947.

Le Beau MM, Rowley JD: Reoccurring chromosomal abnormalities in leukemia and lymphoma. *Cancer Surveys* 1984;3:371–394.

Maniatis T, Fritsch EF, Sambrook J: Molecular cloning: A laboratory manual. Cold Spring Harbor, NY, Cold Spring Harbor Laboratory, 1982.

Melton DA, Krieg PA, Rebagliati MR, Maniatis T, Zinn K, Green MR: Efficient in vitro synthesis of biologically active RNA and RNA hybridization probes from plasmids containing a bacteriophage SP6 promotor. *Nucleic Acids Res* 1984;12:7035–7056.

McGlave P, Scott E, Ramsay N, Arthur D, Blazar B, McCullough J, Kersey J: Unrelated donor bone marrow transplantation therapy for chronic myelogenous leukemia. 1987;70:877–881.

Okayama H, Berg P: A cDNA cloning vector that permits expression of cDNA inserts in mammalian cells. *Mol Cell Biol* 1983;3:280–289.

Pugh WC, Pearson M, Vardiman JW, Rowley JD: Philadelphia chromosome-negative chronic myelogenous leukaemia: A morphological reassessment. *Br J Haematol* 1985;60:457–467.

Punt CJ, Rozenberg-Arska M, Verdonck JF: Successful treatment with chemotherapy and subsequent allogeneic bone marrow transplantation for myeloid blastic crisis of chronic myelogenous leukemia following advanced Hodgkin's disease. *Cancer* 1987;60:934–935.

Ribeiro RC, Abromowitch M, Raimondi SC, Murphy SB, Behm F, Williams DL: Clinical and biologic hallmarks of the Philadelphia chromosome in childhood acute lymphoblastic leukemia. *Blood* 1987; 70:948–953.

Rosenblum MG, Maxwell BL, Talpaz M, Kelleher RJ, McCredie KB, Gutterman JU: In vivo sensitivity and resistance of chronic myelogenous leukemia cells to alpha-interferon: Correlation with receptor binding and induction of 2',5'-oligoadenylate synthetase. Cancer Res 1986;46:4848–4852.

Rowley JD: A new consistent chromosomal abnormality in chronic myelogenous leukaemia identified by quinacrine fluorescence and Giemsa staining. *Nature* 1973;243:290–293.

Secker-Walker LM: The prognostic implications of chromosomal findings in acute lymphoblastic leukemia. *Cancer Genet Cytogenet* 1984;11:233–248.

Shtivelman E, Lifshitz B, Gale RP, Canaani E: Fused transcript of *abl* and *bcr* genes in chronic myelogenous leukaemia. *Nature* 1986;315:550–554.

Southern EM: Detection of specific sequences among DNA fragments separated by gel electrophoresis. *J Mol Biol* 1975;98:503–517.

Stam K, Heisterkamp N, Reynolds FH, Groffen J: Evidence that the ph1 gene encodes a 160,000-dalton phosphoprotein with associated kinase activity. Mol Cell Biol 1987;7:1955–1960.

Stam K, Heisterkamp N, Grosveld G, de Klein A, Verma RS, Coleman M, Dosik H, Groffen J: Evidence of a new chimeric bcr/c-abl mRNA in patients with chronic myelocytic leukemia and the Philadelphia chromosome. *N Engl J Med* 1985;313:1429–1433.

Talpaz M, Kantarjian HM, McCridie KB, Keating MJ, Trujillo J, Gutterman J: Clinical investigation of human alpha interferon in chronic myelogenous leukemia. *Blood* 1987;69:1280–1288.

Talpaz M, Kantarjian HM, McCredie K, Trujillo JM, Keating MJ, Gutterman JU: Hematologic remission and cytogenetic improvement induced by recombinant human interferon alpha A in chronic myelogenous leukemia. *N Engl J Med* 1986;314:1067–1069.

Travis LB, Pierre RV, DeWald GW: Ph1-negative chronic granulocytic leukemia: A nonentity. *Am J Clin Pathol* 1986;85:186–193.

Walker LC, Ganesan TS, Dhut S, Gibbons B, Lister TA, Rothbard J, Young BD: Novel chiameric protein expressed in Philadelphia positive acute lymphoblastic leukaemia. *Nature* 1987;329:851–853.

Yoffe G, Blick M, Kantarjian H, Spitzer G, Gutterman J, Talpaz M: Molecular analysis of interferon-induced suppression in chronic myelogenous leukemia. *Blood* 1987;69:961–963.

Yunis JJ, Brunning RD, Howe RB, Lobell M: High-resolution chromosomes as an independent prognostic indicator in adult acute nonlymphocytic leukemia. *N Engl J Med* 1984;311:812–818.

Yuo A, Kitagawa S, Okabe T, Urabe A, Komatsu Y, Itoh S, Takaku F: Recombinant human granulocyte colony-stimulating factor repairs the abnormalities of neutrophils in patients with myelodysplastic syndromes and chronic myelogenous leukemia. *Blood* 1987;70:404–411.

Karyotype Interpretation

7.1 HISTORICAL ACCOUNT

The 44 autosomes and two sex chromosomes of the human genome are unique. However, various chromosomes resemble each other when stained by conventional methods, obscuring individual identification. Chromosomes were initially classified into seven groups by size (A-G). The X-chromosome was placed within the C-group, whereas the Y chromosome was placed in the G-group. Although identification of chromosomes by group was a significant achievement, most of the extra or structurally altered chromosomes could not be identified. Several investigators in the early 1960s attempted to identify individual chromosomes by autoradiographic techniques. However, neither chromosome morphology nor autoradiography provided unequivocal identification.

In the early 1960s a handful of investigators met in Denver to discuss the proposed nomenclature for interpreting and reporting karyotypes to clinicians (Denver Conference, 1960). Three years later another meeting was called to include the developments that had occurred since the Denver Conference (London Conference, 1963). The major achievement of this conference was to officially recognize the seven groups originally proposed by Patau (1960). The need to improve karyotypic classification and nomenclature that continued to puzzle many cytogeneticists and clinicians. In 1966 they assembled in Chicago to revise the nomenclature once again to include the rapid developments that had occurred since 1963 (Chicago Conference, 1966). The major breakthrough in chromosomal identification occurred after Caspersson's group (1971) demonstrated that each chromosome has its own anatomy by virtue of its banding pattern. Thus, it was declared that the existing system was no longer sufficient to describe chromosomal abnormalities found in humans. With the improved methodology each of us realized that a deeper and finer classification of each band was possible. The fourth international congress on human genetics held in Paris was a landmark achievement, for it was now possible to identify chromosomes by regions and bands. Furthermore, it paved the way to describe chromosomal changes that were acquired or inherited.

The standing committee met in Stockholm in 1977, reviewed the earlier proposed system, and compiled a single document entitled, "An International System for Human Cytogenetic Nomenclature" (ISCN, 1978). A number of laboratories began to initiate cultures with partial synchronization using various chemicals to obtain prophase and prometaphase chromosomes. The beadlike structures seen on these longer chromosomes gained recognition as so-called high resolution bands; subclassification of bands was inevitable. Cytogenetic nomenclature for these high resolution bands was devised by a group of investigators who met in Paris in 1980. An official publication, entitled "An International System For Human Cytogenetic Nomenclature – High Resolution Banding" (1981), was produced. This publication has become widely accepted by the scientific community and proven to be the definitive source for communication between fellow scientists. Human chromosomes are numbered from 1 to 22, X and Y. Classification of chromosomes is based on the size and arm ratios.

7.2   CLASSIFICATION OF UNBANDED CHROMOSOMES

The following description was recommended for individual chromosomes by the standing committee on nomenclature (ISCN, 1985):

Group (A) Chromosomes 1–3	Large metacentric chromosomes readily distinguished from each other by size and centromere position.
Group (B) Chromosomes 4–5	Large submetacentric chromosomes which are difficult to distinguish from each other.
Group (C) Chromosomes 6–12	Medium-sized metacentric chromosomes. The X chromosome resembles the longer chromosomes in this group. This large group is the one which presents major difficulties in identification of individual chromosomes without the use of banding techniques.
Group (D) Chromosomes 13–15	Medium-sized acrocentric chromosomes with satelites.
Group (E) Chromosomes 16–18	Relatively short metacentric chromosomes (no. 16) or submetacentric chromosomes (no. 17 and 18).
Group (F) Chromosomes 19–20	Short metacentric chromosomes.
Group (G) Chromosomes 21–22–Y	Short acrocentric chromosomes with satellites; the Y chromosome is similar to these chromosomes but bears no satellites.

Not all chromosomes in the D and G groups always exhibit satellites. Also, there may be some length variation between homologues of chromosomes 1, 9, and 16 due to the secondary constriction region (h). Detailed description of such polymorphic variations has been given elsewhere (Babu and Verma, 1987).

7.3   NOMENCLATURE OF THE BANDED
MITOTIC CHROMOSOMES

We decided to include introductory descriptions of the nomenclature for banded chromosomes because of the diversity of background knowledge of our readers. The most commonly used term is "band," defined as part of the chromosome clearly distinguishable from its adjacent segment by its differential staining property. The longitudinal differentiation of individual chromosomes displayed by a continuous series of bands remains one of the most intriguing properties of chromosomes today. Because of the specific landmarks of the short (p) and long arms (q), each arm was first divided by regions. The bands and regions are numbered from the centromere outward. A region is defined as any area of a chromosome lying between two adjacent landmarks. The regions adjacent to the centromere are labeled #1 in each arm, whereas the adjacent distal region is #2 and so on. When identifying a particular area of a chromosome, four criteria are involved: (1) the chromosome number, (2) the arm symbol, (3) the region number, and (4) the band number within that region.

The diagrammatic representation of bands was first established at the Paris Conference (1971) and was based on the patterns observed in different cells stained with either the Q-, G-, or R-banded techniques. This diagram was sufficient to provide nomenclature. The position of these bands, however, was not based on measurements or sequentially banded cells, but was designated solely on the basis of their midpoints. In general, whenever an existing band is subdivided, a decimal point is placed after the original band designation followed by the number assigned to each subband. Again, like bands, the subbands are also numbered sequentially from the centromere outward. The variations in secondary constriction regions (h) of chromosomes 1, 9, and 16 and satellites (s) on chromosomes 13, 14, 15, 21, and 22 are distinguished from increases and decreases in arm length by placing symbol h or s between the symbol for the arm and the plus or minus sign. For example, 46,XY,16qh+ translates to a male karyotype with 46 chromosomes, showing an increase in length of the secondary constriction on the long arm of chromosome 16.

A series of symbols were proposed to describe normal and aberrant chromosomes (Table 7.3.1). In describing a karyotype, the first item to be recorded is the total number of chromosomes followed by a comma (,). The sex chromosome constitution is given next. It would be redundant here to describe the nomenclature for various types of chromosomal abnormalities. Detailed description is given elsewhere (ISCN, 1985) and is readily available by writing to S. Karger AG, P.O. Box, CH-4009, Basel, Switzerland. However, a brief description of the most common abnormalities follows. These examples are borrowed from ISCN (1985).

Table 7.3.1. Most Commonly Used Symbols and Abbreviations in Cytogenetic Nomenclature*

Symbols	Description
ace	Acentric fragment
arrow (→)	From → to
b	Break
cen	Centromere
colon, single (:)	Break
colon, double (::)	Break and reunion
cs	Chromosome
ct	Chromatid
del	Deletion
der	Derivative chromosome
dup	Duplication
end	Endoreduplication
f	Fragment
fra	Fragile site
g	Gap
h	Secondary constriction
i	Isochromosome
ins	Insertion
inv	Inversion
mar	Marker chromosome
mat	Maternal origin
minus (−)	Loss
mos	Mosaic
p	Short arm of chromosome
pat	Paternal origin
Ph	Philadelphia chromosome
plus (+)	Gain
q	Long arm of chromosome
qr	Quadriradial
r	Ring chromosome
rcp	Reciprocal
rea	Rearrangement
rec	Recombinant chromosome
rob	Robertsonian translocation
s	Satellite
sce	Sister chromatid exchange
semicolon (;)	Separates chromosomes and chromosome regions in structural rearrangements involving more than one chromosome
t	Translocation
tan	Tandem translocation
ter	Terminal (end of chromosome)
tr	Triradial
var	Variable chromosome region

*Annotated from ISCN (1985).

Terminal Deletions

46,XX,del(1)(q21)

46,XX,del(1)(pter→q21:)

The single colon (:) indicates a break at band 1q21 and deletion of the long-arm segment distal to it. The remaining chromosome consists of the entire short arm of chromosome 1 and part of the long arm lying between the centromere and band 1q21.

Interstitial Deletions

46,XX,del(1)(q21q31)

46,XX,del(1)(pter→q21::q31→qter)

The double colon (::) indicates breakage and reunion of bands 1q21 and 1q31 in the long arm of chromosome 1. The segment lying between these bands has been deleted.

Paracentric Inversions

46,XY,inv(2)(p13p24)

46,XY,inv(2)(pter→p24::p13→p24::p13→qter)

Breakage and reunion have occurred at bands 2p13 and 2p24 in the short arm of chromosome 2. The segment lying between these bands is still present but inverted, as indicated by the reverse order of the bands with respect to the centromere in this segment of the rearranged chromosome.

Pericentric Inversions

46,XY,inv(2)(p21q31)

46,XY,inv(2)(pter→p21::q31→p21::q31→qter)

Breakage and reunion have occurred at band 2p21 in the short arm and 2q31 in the long arm of chromosome 2. The segment lying between these bands is inverted.

Isochromosomes

46,X,i(Xq)

46,X,i(X)(qter→cen→qter)

Breakpoints in this type of rearrangement are at or close to the centromere and cannot be specified. The designation indicates that both entire long arms of the X chromosome are present and are separated by the centromere.

Ring Chromosome

46,XY,r(2)(p21q31)

46,XY,r(2)(p21→q31)

Breakage has occurred at band 2p21 in the short arm and 2q31 in the long arm of chromosome 2. With deletion of the segments distal to these bands, the broken ends have joined to form a ring chromosome. Note the omission of the colon or double colon.

Dicentric Chromosomes

46,X,dic(Y)(q21)

46,X,dic(Y)(pter→q21::q12→pter)

Breakage and reunion have occurred at band Yq12 on sister chromatids to form a dicentric Y chromosome. The alternative form of a dicentric Y chromosome might be:

46,X,dic(Y)(p11)

46,X,dic(Y)(qter→p11::p11→qter)

Fragile Site

46,XY,fra(X)(q27)

Male with a fragile site at band Xq27, seen in the Marker X or Fragile X syndrome.

Reciprocal Translocations

46,XY,t(2;5)(q21;q31)

46,XY,t(2;5)(2pter→2q21::5q31→5qter;5pter→5q31::2q21→2qter)

Breakage and reunion have occurred at bands 2q21 and 5q31 in the long arms of chromosomes 2 and 5, respectively. The segments distal to these bands have been exchanged between two chromosomes. The derivative chromosome with the lowest number (ie, chromosome 2) is designated first.

7.4   HIGH RESOLUTION BANDS OF CHROMOSOMES

Improved technical achievements in banding chromosomes has added a new dimension to increasing band resolution. It is a well-documented fact that prophase and prometaphase chromosomes reveal far more bands than can be seen even in the best banded metaphase chromosomes. This in turn has provided a greater resolution for precise identification of chromosomal abnormalities. Synchronization of cultures by methotrexate, which delays DNA synthesis and subsequent thymidine release, has been an original approach to obtaining earlier stages of cell division (Yunis, 1976). Since 1976, many cell synchronization techniques have been developed. In an earlier chapter we described this procedure under the heading of high resolution banding techniques. It is a well proven fact that this approach has provided a more accurate diagnosis of chromosomal defects in humans. Therefore, the need for a standard system of nomenclature for the increased number of bands of the human genome was created. The working committee met several times and a remarkable agreement was reached in May 1980.

The seven principles on which the nomenclature is based are as follows:

1. Nomenclature for high resolution bands are an extension of the existing nomenclature proposed at the Paris Conference (1971). Minor changes are made only when it becomes an absolute necessity.

2. Different bands are seen at different stages of mitosis. The diagram shown in Figure 7.4.1a provides schematic representation of chromosomes corresponding to approximately 400, 550, and 850 bands.

3. Bands vary in width and intensity depending on which technique is used. Thus, a simple black and white idiogram was drawn to indicate position and relative width only. The primary and secondary constriction regions indicated by crosshatching are the only exceptions to this rule.

(Text continues on p. 205)

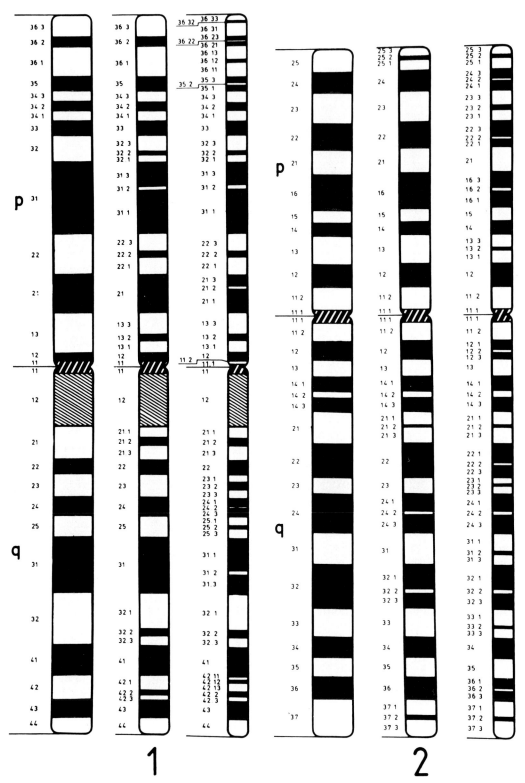

FIGURE 7.4.1. Diagrammatic representation of G-banded chromosomes at three different resolutions. The left chromosome represents a haploid karotype at approximately the 400 band level. The central figure represents a karyotype of approximately 550 bands. The diagram on the far right represents a karyotype of approximately 850 bands. (Reproduced, with permission, from S. Karger AG, Basel.) *Figure continued on the following pages.*

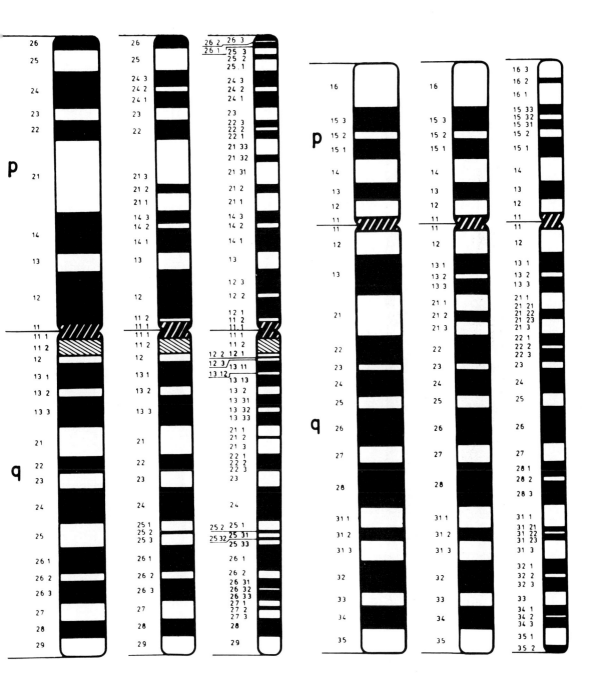

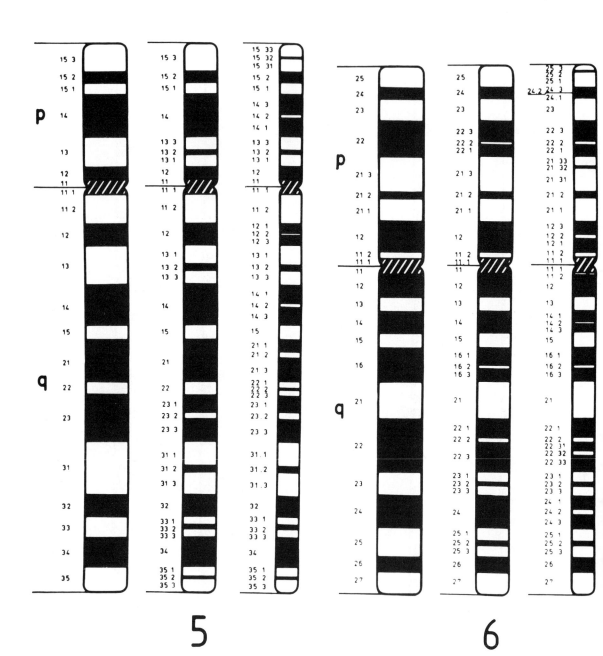

FIGURE 7.4.1 continued.

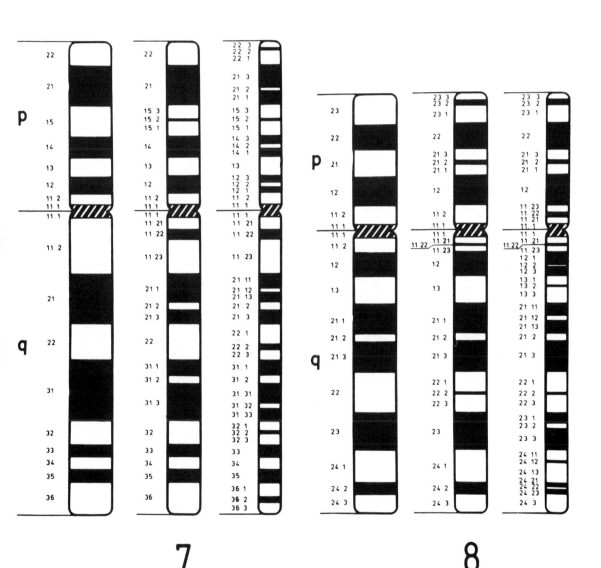

FIGURE 7.4.1 continued.

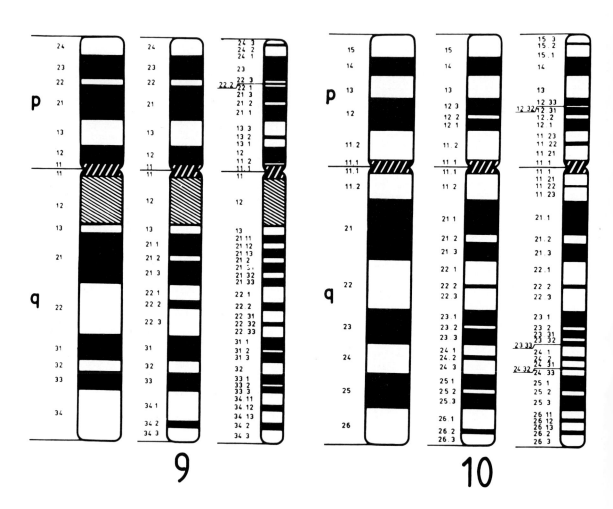

FIGURE 7.4.1 continued.

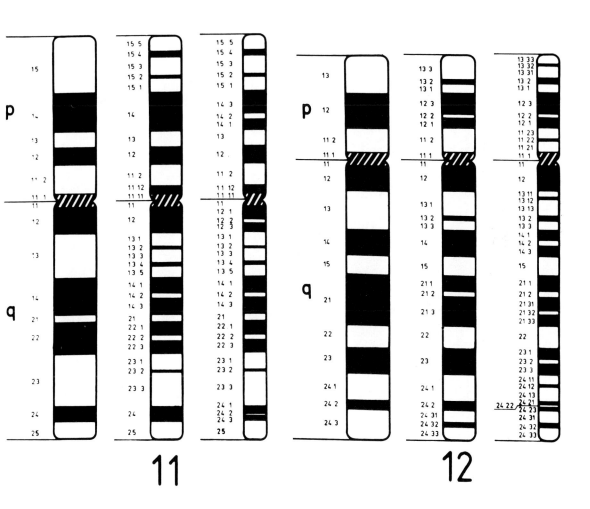

FIGURE 7.4.1 continued.

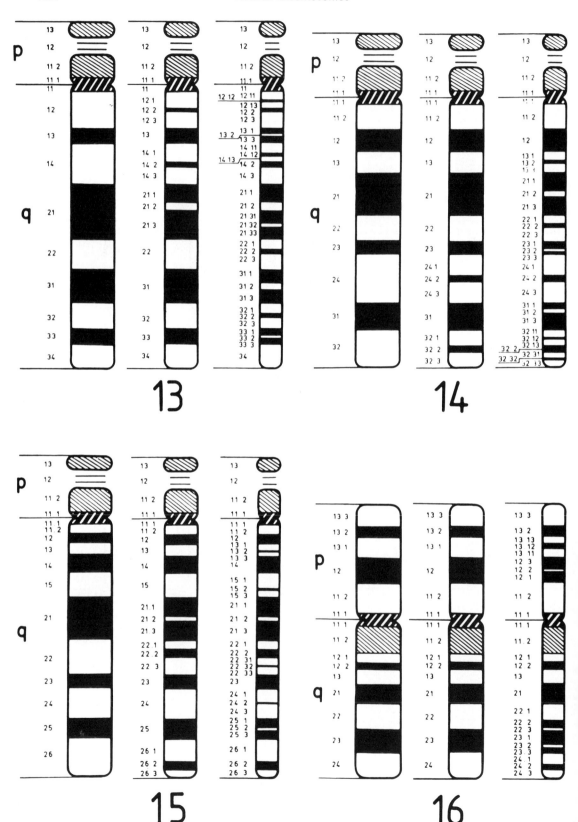

FIGURE 7.4.1 continued.

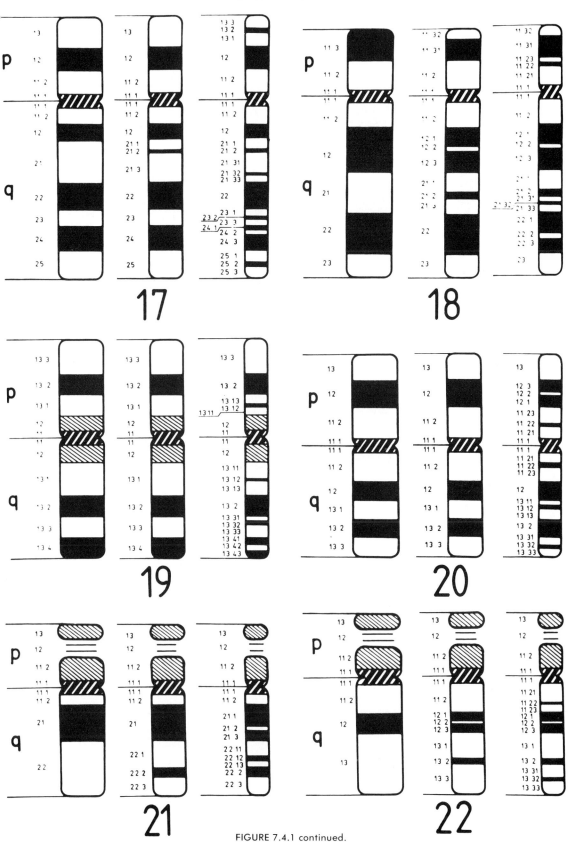

FIGURE 7.4.1 continued.

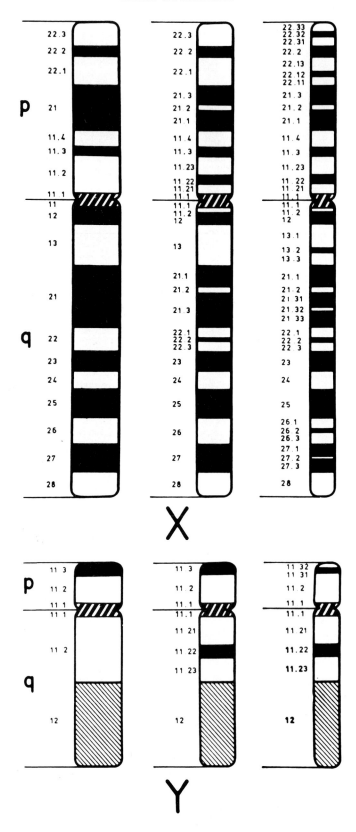

FIGURE 7.4.1 continued.

4. Positions of the bands are relative.

5. There are no interrelationships between bands produced by various techniques. Thus, to avoid confusion, G-banding patterns were given prime importance for numbering.

6. The demarcation of bands was arbitrary because band physiology in different arms still remains intriguing. Detailed deviations of individual bands from the Paris (1971) and ISCN (1978) diagrams are described elsewhere (ISCN, 1981).

7. The nomenclature remains the same as described in ISCN (1978). The basic rule of thumb is to add a new digit when subdividing bands. For example, band 3q25 was subdivided into 3q25.1, 3q25.2, and 3q25.3; subsequently, 3q25.3 can be subdivided into 3q25.31, 3q25.32, 3q25.33, and so forth.

7.5   CHROMOSOMAL ABERRATIONS

Chromosomal breakage occurs at various stages of mitosis (ie, G_1, S, G_2, or M) and during meiosis. Therman (1980) and Auerbach (1976) provide comprehensive reviews of this subject. If breaks occur during the G_1-stage, only one chromatid is affected. Both chromatids are involved if breakage takes place during the S phase. When two breaks occur in the same chromosome, centric rings or acentric fragments are produced. There are certain autosomal diseases that show a high frequency of spontaneous chromosomal aberrations or a high susceptibility to induced damage. The method and purpose of scoring these aberrations depend on the purpose for which the data are required. Maximum information can only be obtained when data are classified by the types of chromosomal aberrations. Buckton and Evans (1973) clearly provided an extensive protocol for classification of these chromosomes which was annotated by Savage (1976). The nomenclature for these acquired aberrations has been adequately described by the existing terminology. Therefore, we believe it is redundant to discuss the matter here.

7.6   NOMENCLATURE FOR HUMAN MEIOTIC CHROMOSOMES

Meiotic chromosomes in humans have not become popular because preparation is time-consuming. Most investigations are limited to a basic understanding of meiosis. Nevertheless, analysis of meiotic chromosomes is essential for risk assessment of genetic carriers and their relatives. Therefore, we decided to devote a section to this technique as well (Section 2.7). In general, Q-, G-, and C-banding has been applied to first prophase through first metaphase chromosomes. Autosomal bivalents generally show the same banding patterns as seen in somatic metaphases; however, some minor differences are found between the bivalents and mitotic chromosomes with C-banding. The relative length, centromeric index, and number of chiasmata per bivalent have been worked out. The abbreviations PI, MI, and the like are used to designate prophase I, metaphase I, and so forth, followed by the total count of separate chromosomal elements. As usual, the sex chromosomes are then indicated by XY, or XX; for example, MI, 23, XY is a primary spermatocyte at diakinesis or metaphase I with 23 elements including an XY bivalent.

A detailed description of banding patterns obtained at pachytene using 850-stage band nomenclature of somatic chromosomes is described in ISCN (1985) and reproduced in Figure 7.6.1. Chromomeres, unique meiotic structures, are assigned as Giemsa-positive bands, whereas interchromomere regions are numbered as Giemsa-negative bands. The pufflike structure located at the heterochromatic region of chromosome 9 band q12 is a unique feature of human pachytene chromosomes.

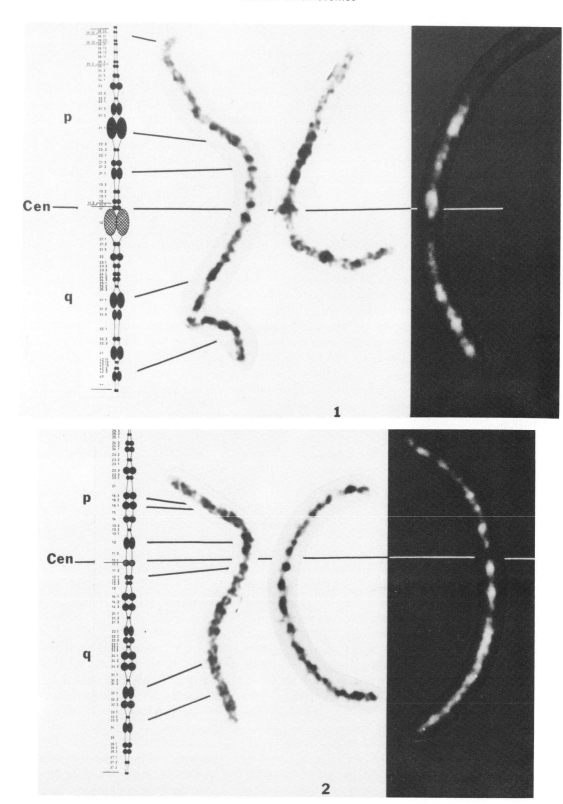

FIGURE 7.6.1. Chromomere idiogram of the 22 autosomal bivalents at pachytene. The resolution of the bands is at approximately the 850-band level. The nomenclature remains the same as that described for somatic chromosomes in Figure 7.4.1. (Courtesy of Drs. R. S. K. Chaganti and S. C. Jhanwar; reproduced, with permission, from S. Karger AG, Basel.) *Figure continued on the following pages.*

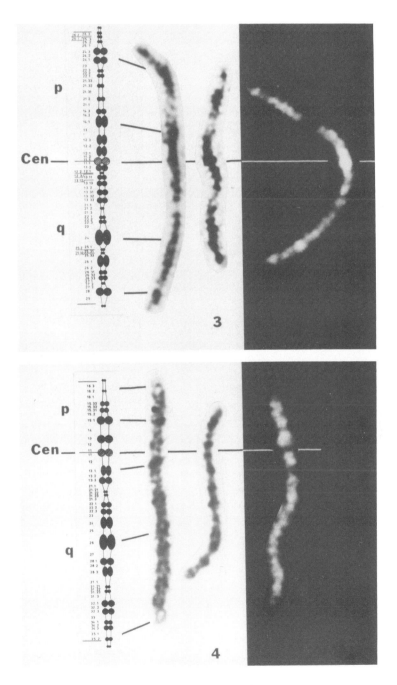

FIGURE 7.6.1 continued.

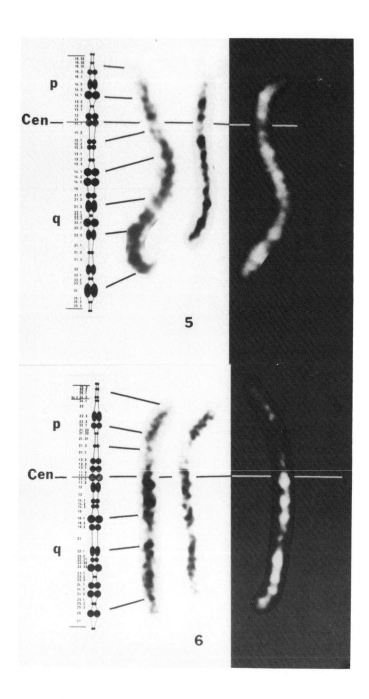

FIGURE 7.6.1 continued.

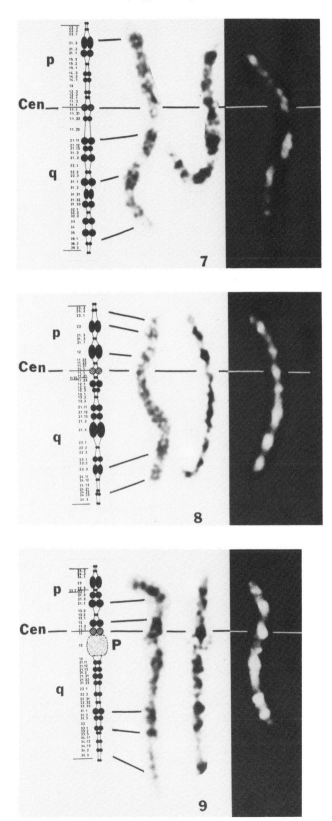

FIGURE 7.6.1 continued.

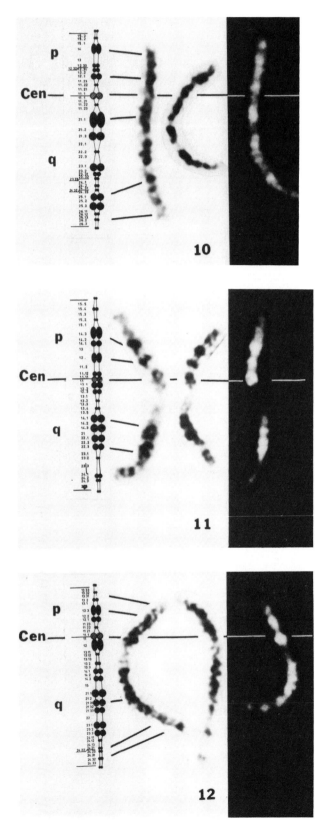

FIGURE 7.6.1 continued.

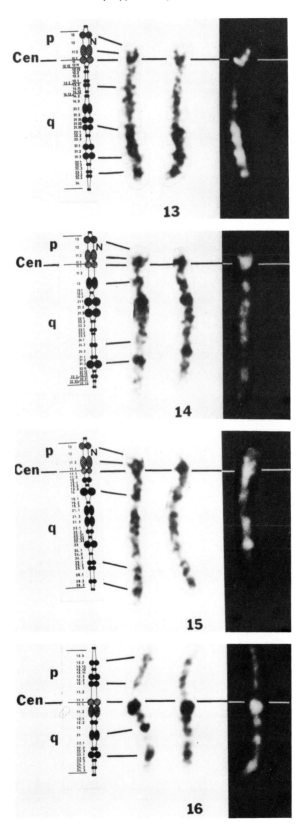

FIGURE 7.6.1 continued.

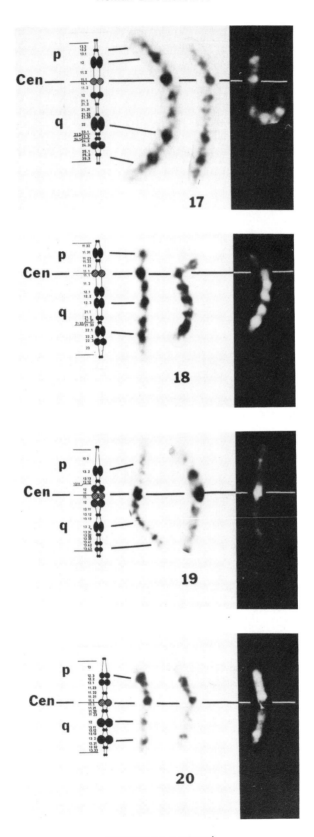

FIGURE 7.6.1 continued.

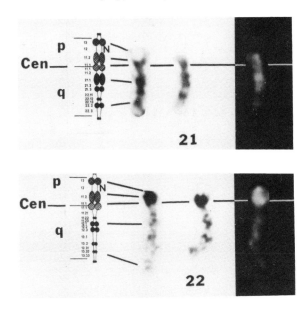

FIGURE 7.6.1 continued.

7.7   REFERENCES

Auerbach C: *Mutation Research. Problem Results and Perspectives.* London, Chapman and Hall, 1976.

Babu A, Verma RS: Chromosome structure. Euchromatin and heterochromatin, in Bourne GH, Jeon KW, Friedlander (eds): *International Review of Cytology.* New York, Academic Press, 1987, vol. 108, pp 1–60.

Buckton KE, Evans HJ (ed): *Methods for the Analysis of Human Chromosome Aberrations.* Geneva, World Health Organization, 1973.

Caspersson T, Lomakka G, Zech L: The 24 fluorescence patterns of human metaphase chromosomes – distinguishing characters and variability. *Hereditas* 1971;67:89–102.

Chicago Conference (1966): *Standardization in Human Cytogenetics: Birth Defects.* Original Article Series, Vol 2, No 2. New York, The National Foundation, 1966.

Denver Conference (1960): A proposed standard system of nomenclature of human mitotic chromosomes. *Lancet* 1960;I:1063–1065.

ISCN (1978): *An International System for Human Cytogenetic Nomenclature: Birth Defects.* Original Article Series, Vol 14, No 8. New York, The National Foundation, 1978; also in *Cytogenet Cell Genet* 1978;21:309–404.

ISCN (1981): *An International System for Human Cytogenetic Nomenclature – High Resolution Banding (1981): Birth Defects.* Original Article Series, Vol 17. New York, March of Dimes Birth Defects Foundation, 1981; also in *Cytogenet Cell Genet* 1981;31:1–23.

ISCN (1985): *An International System for Human Cytogenetic Nomenclature: Birth Defects.* Original Article Series, Vol 21, No 1. New York, The National Foundation, 1985.

London Conference on the Normal Human Karyotype (1963): *Cytogenetics* 1963;2:264–268. Reprinted in Chicago Conference, 1966, pp 18–19.

Paris Conference (1971): *Standardization in Human Cytogenetics: Birth Defects.* Original Article Series, Vol 8, No 7, New York, The National Foundation, 1972; also in *Cytogenetics* 1972;11:313–362.

Paris Conference (1971), Supplement (1975): *Standardization in Human Cytogenetics: Birth Defects.* Original Article Series, Vol 11, No 9, New York, The National Foundation, 1975; also in *Cytogenet Cell Genet* 1975;15:201–238.

Patau K: The identification of individual chromosomes, especially in man. *Am J Hum Genet* 1960; 12:250–276.

Savage JRK: Annotation: Classification and relationships of induced chromosomal structural changes. *J Med Genet* 1976;13:103–122.

Therman E: *Human Chromosomes: Structure, Behavior, Effects.* New York, Springer-Verlag, 1980.

Yunis JJ: High resolution of human chromosome. *Science* 1976;191:1268–1270.

Yunis JJ, Sawyer JR, Ball DW: The characterisation of high resolution G banded chromosomes of man. *Chromosoma* 1978;67:293–307.

8

Cell Cultures: Maintenance and Storage

Laboratories frequently encounter some unusual and interesting cases that need elaborate work using a series of different protocols for either diagnosis or research purposes. It is necessary to maintain either lymphoblastic or fibroblastic cells for a period of time through several subcultures and to store for future use. The established cell lines can be maintained using simple protocols. However, establishing lymphoid cell lines and maintaining and storing primary fibroblast cell lines require special attention and manipulations. Maintaining the cells in culture for long periods, when not used for a particular work, is time-consuming and uneconomical. Moreover, there is a constant threat of losing valuable cells by either the natural aging process, karyotypic changes, or contamination. Elaborate technical information can be found in some textbooks (Barnes et al, 1984; Freshney, 1983; Kruse and Patterson, 1973). Some protocols for maintenance and storage of cell cultures helpful in day-to-day work are described in the following section.

Cell cultures have two basic needs: (1) Cells need to be fed by removing the exhausted medium and replenishing it with fresh medium at regular intervals to maintain the required levels of nutrients, and (2) the cultures need to be divided (subculture) to maintain an optimum cell population that would sustain active cell proliferation.

8.1 MAINTAINING SUSPENSION CULTURES

Cell cultures of lymphoid origin generally grow as free cells in medium without anchoring to the substrate. A number of human cell cultures are of this type. These cell lines are being established by an increasing number of laboratories that use lymphocytes from peripheral blood samples.

In addition, some established cell lines of fibroblastic origin are maintained and grown in suspension cultures by mechanical devices that either constantly rotate or shake the culture vessels to prevent the cells from adhering to any surface. The later methods are adopted to meet special needs in obtaining large volumes of cells, but they are usually not required for a cytogenetic laboratory. Therefore, these techniques are ignored here.

Protocol 8.1.1   Feeding the Suspension Cultures

Solutions

1. Growth medium

MEM for suspension cultures	100 ml
Fetal bovine serum (FBS)	20 ml
Penicillin and streptomycin solution	1.3 ml
(10,000 U/ml and 10,000 μg/ml, respectively)	

A number of other formulated media can be substituted for MEM for suspension cultures. The amount of FBS can be reduced to 5 or 10% because the established lymphoblast cell lines can grow equally well in media with low serum content. Growth medium can be stored at 2 to 5°C for 2 to 3 weeks.

Procedure

1. Suspension cultures of lymphoblasts are conveniently maintained in T-type (T25) flasks incubated in vertical position with about 20 to 30 ml of medium per flask.
2. Transfer the flasks to a working sterile area (laminar flow hood) without disturbing the cells settled at the bottom of the flask.
3. Remove aseptically 10 to 15 ml of supernatant-cell-free medium.
4. Replenish the flask with an equal amount of fresh growth medium and suspend the cells by a gentle swirling motion.
5. Return the flasks to the incubator and incubate the culture flasks in a vertical position. (If incubating in a CO_2 incubator, loosen the caps slightly to allow the exchange of gases.)

Comments

If for any reason the suspension cultures are to be incubated in a horizontal position, the amount of medium should be reduced to prevent spillage or incubated as closed cultures with the cap tightened. These cultures would require additional attention to maintain optimum pH of medium.

Medium is partially renewed in the cultures in the protocol just described. Alternatively, a complete medium change can be performed by centrifuging the contents of the flask, discarding the old medium, and replenishing it with fresh growth medium.

Protocol 8.1.2   Subculturing the Suspension Cultures

Solutions

1. Growth medium
   As just described.

Procedure

1. Transfer the culture flasks, maintaining the vertical position without disturbing the cells settled at the bottom of the flask.
2. Remove the supernatant-clear medium from the culture flask, leaving about 10 ml of medium.
3. Suspend the cells in the leftover medium by gently swirling the flask.
4. Obtain a small volume of suspension for cell counting and count the cells using a hemocytometer or an electronic cell counter.
5. Add fresh growth medium to adjust the cell density (1×10^6 cells per milliliter of medium).
6. Distribute 25 to 30 ml of cell suspension into fresh culture flasks as needed.
7. Transfer the cultures to an incubator.
8. Feed the cultures on alternate days or as needed depending on the rate of cell proliferation following protocol 8.1.1.

Comments

The flasks are subcultured once a week or every 10 days. Cultures with very active growth may require more frequent subculturing. Cell counting and adjustment to the final cell density (1×10^6 cells per milliliter of medium) are suggested in subculturing for theoretic reasons. However, for routine maintenance of a cell line for cytogenetic purposes, cell counting is neither crucial nor required. A split ratio of 1:4 or 1:5 usually serves the purpose for a number of lymphoid cell lines. The split ratio can be adjusted to meet a specific cell line or subculture interval.

8.2   MAINTAINING THE MONOLAYER CULTURES

Most of the fibroblastic or epithelial cell lines grow in a monolayer by adhering to the surface of the culture vessels (usually the T-type flasks). The culture flasks are maintained in a horizontal position to provide a large surface area to which cells can adhere. The cell lines derived from a number of tissues, such as amniotic fluids, skin, kidney, and lung, are usually fibroblasts. Renewing the growth medium in these cultures is relatively simple, but subculturing requires dislodging the cells from the culture flask with a trypsin solution and washing the cells to remove the enzyme.

Protocol 8.2.1   Feeding the Monolayer Cultures

Solutions

1. Growth medium

RPMI 1640	100 ml
FBS	20 ml
Penicillin and streptomycin solution	1.3 ml
(10,000 U/ml and 10,000 μg/ml, respectively)	

A number of other formulated media can be used instead of RPMI 1640 to prepare growth medium.

Procedure

1. The monolayer cultures grown in T-type flasks are incubated in a horizontal position in a CO_2 incubator with loose caps. Tighten the caps and transfer the flasks to the work area.
2. Following sterile techniques, discard the old medium (as much as possible) and add an appropriate amount of fresh growth medium, depending on the size of the flask (usually 5 ml per T25 flask).
3. Transfer the cultures back into the CO_2 incubator and incubate in a horizontal position. Loosen the caps by a half turn.

Comments

Maintain aseptic conditions throughout the work, and care should be taken to prevent cross-contamination of cells between cell lines. The cell layers should not be left for any period without bathing in the medium to avoid even partial drying of the cells. The culture flasks are better if maintained in a horizontal position during the transfer and working periods.

Protocol 8.2.2   Subculturing the Monolayer Cultures

Solutions

1. Growth medium
   Same as in protocol 8.2.1.

2. Calcium- and magnesium-free phosphate-buffered saline (CMS-PBS) with trypsin EDTA.

This solution is commercially available for tissue culture purposes at 1X and 10X concentrations. It is used at 1X concentration. The concentrations of trypsin (1:250) and EDTA in 1X solution are 0.25 and 0.038%, respectively.

Procedure

1. Tighten the caps and transfer the flasks to the work area.

2. Discard as much old medium as possible from the flask.

3. Place a small amount of CMS-PBS (5 ml for T25 flask or more for larger flasks) and thoroughly rinse the entire monolayer by tilting the flask from side to side. Discard the solution. (This rinsing step is intended to dilute and remove traces of leftover growth medium that contains serum. Trypsin-EDTA solution can also be used instead of CMS-PBS for rinsing. In this case the volume should be small, about 1 ml for a T25 flask or more for larger flasks. Rinsing should be brief. Prolonged rinsing may remove some of the cells that would be lost.)

4. Add a small amount of fresh trypsin-EDTA solution (1.5 ml for T25 or more for larger flasks). Bathe the cells evenly by tilting the flask from side to side and incubate at 37°C for 5 minutes.

5. Examine the flasks following the incubation period. The cell layers are usually detached from the substrate during this period. If the cells are not detached, continue incubation for a few additional minutes.

6. Add 8.5 ml of fresh growth medium to the flask and suspend the cells uniformly by passing the contents through a pipette a few times. Perform cell counting. Adjust the final cell density to approximately 1×10^6 cells per milliliter.

7. Distribute the cell suspension at appropriate volumes to new culture flasks.

8. Incubate the cultures in a CO_2 incubator for 2 to 3 hours.

9. Discard the medium containing trypsin-EDTA and replenish it with fresh growth medium.

10. Feed the cultures on alternate days or as required.

Comments

Cells should not be kept in serum-free medium (in this case trypsin-EDTA solution) for more than 10 to 15 minutes to prevent permanent cell injury. The procedure just described eliminates the need for centrifugation and thus prevents cell loss during transfers back and forth from the culture tubes. The enzymatic activity of trypsin is diluted and inhibited by serum in growth medium. However, the medium must be renewed after 2 or 3 hours to remove EDTA from the cultures. Viable cells usually adhere to the growth surface within 40 to 60 minutes.

Alternatively, the trypsinized cell suspension (step 6) can be centrifuged and resuspended in 10 ml of fresh growth medium before cell counting. In this case, the medium renewal (step 9) should be omitted.

As mentioned earlier for suspension cultures, cell counting and adjusting the cell density are not essential. The monolayer cultures are subcultured before reaching confluency. The split ratio and subculture interval vary, depending on the growth rate of a particular cell line. Established cell lines generally require higher split ratios of 1:5 or more. It is preferred to maintain lower split ratios (about 1:2 or 1:3) for primary and secondary cell cultures and gradually increase them as the number of subcultures is increased.

8.3 CELL STORAGE AND RECOVERY

Tissue culture cells can be stored at cold temperatures for long periods (cryopreservation) by protecting them from the damage caused by freezing with the help of glycerol or dimethyl-sulfoxide (DMSO) which are commonly known as cryoprotective agents. Cell storage is now a common practice that saves time and money and above all the valuable cell lines that could otherwise be lost. Cells are stored at low temperatures in either mechanical refrigerators (−90°C), dry ice (−79°C), or liquid nitrogen (−196°C). Although the cells survive equally well at all these temperatures for an initial period of several months, the number of viable cells rapidly decreases over the years when stored in dry ice. Conversely, cells stored in liquid nitrogen show consistently high survival rates over the years.

Protocol 8.3.1 Cell Storage

Solutions

1. Preservation medium

Growth medium	95 ml
Glycerol	5 ml

   The concentration of glycerol can be increased up to 10 ml. However, a low concentration of cryoprotective agents has the advantage of eliminating the need to wash the cells during recovery. Glycerol can be substituted for DMSO at an equal quantity. The cryoprotective agents are sterilized before adding them to the growth medium. Glycerol is sterilized in small aliquots by autoclaving. DMSO is sterilized by filtering through a Selas filter candle with an 0.03 μ pore size and stored in small aliquots at 2 to 5°C.

2. Trypsin-EDTA solution
   Same as in protocol 8.2.1.

Procedure

1. Obtain an exponentially growing cell culture.
2. For monolayer culture, the cells should be trypsinized (steps 1 to 5 in protocol 8.2.2) and suspended by adding 5 to 10 ml of fresh growth medium. For suspension cultures, trypsinization is not required. Instead, bring the cells into suspension and follow the next steps.
3. Transfer the cell suspension into centrifuge tubes (cells of monolayer in T25 flask/tube or 10 ml of cell suspension of suspension culture/tube) and centrifuge at 800 rpm for 10 minutes. Discard the supernatant medium.
4. Suspend the cells of each centrifuge tube in 1 ml of preservation medium.
5. Transfer the suspension to preservation ampules and close the vials tightly. Label each vial with proper information regarding the cells. Use appropriate labeling material; stick-on labels may come off.
6. Cool the vials gradually by decreasing the temperature 1° to 2°C per minute. This is accomplished by refrigerating the ampules for 1 hour, transferring them into the gaseous phase of a liquid nitrogen tank and leaving them overnight. Finally, lower the ampules into the liquid nitrogen.
7. Place the ampules on canes and store. Log the information about the cells stored including details of the cell line, the number of vials, and the number of canes.

Comments

The cooling of storage vials should be gradual, a rate of approximately 1°C per minute. The use of plastic ampules avoids the danger of explosion during the recovery process. Use protective gloves when handling the material to avoid cold burns.

Protocol 8.3.2   Cell Recovery

Solutions

1. Growth medium
   Use appropriate growth medium as described in protocol 8.1.1 or 8.2.1 for suspension culture or monolayer culture.

Procedure

1. Wearing protective gloves, remove the ampule(s) from storage, immediately plunge it into a water bath at 37°C, and agitate the contents of the vials as they thaw.

2. Immerse the ampule(s) in 70% ethanol.

3. Transfer the contents of an ampule into a culture flask. Add 9 ml of growth medium and incubate in a CO_2 incubator. Change the medium the following day. Washing the cells is not required if 5% of cryoprotective agent is used.

or

Transfer the contents of an ampule into a centrifuge tube and add 10 ml of growth medium. Centrifuge at 800 rpm for 10 minutes and discard the supernatant. Suspend the cells in 5 ml of growth medium, transfer the suspension into a culture flask, and incubate in a CO_2 incubator.

4. Feed the cells and subculture according to routine protocols.

Comments

The thawing process should be as rapid as possible. If glass ampules are used for storage, a protective mask should be worn because of the risk of explosion.

8.4   REFERENCES

Barnes DW, Sirbasku DA, Sato GH (eds): *Cell Culture Methods for Molecular and Cell Biology* (4 Volumes). New York, Alan R. Liss, Inc., 1984.

Freshney RI: *Culture of Animal Cells. A Manual of Basic Techniques.* New York, Alan R. Liss, Inc., 1983.

Kruse PF Jr, Patterson MK: *Tissue Culture. Methods and Application.* New York, Academic Press, 1973.

9

Chromosomes in Clinical Medicine

The invention of banding procedures coupled with molecular cytogenetics has permitted knowledge of the role of chromosomes in clinical medicine to flourish. These technical advances, together with high resolution banding techniques, allow precise dissection of chromosomes, leading to localization of defective genes in the human genome. The significance of these achievements is clearly reflected in the discovery of an ever-increasing number of sites responsible for human diseases. Furthermore, the isolation of satellite DNA and the use of restriction endonucleases have elucidated classes of DNA responsible for nondisjunction, producing euploidy and aneuploidy.

Information within the human genome has been unmasked by the recently developed technical tools that are covered in various chapters. New methods have led to bursts of knowledge about human chromosomes. The great triumphs possible with staining techniques have led to the creation of a new subspecialty, called *clinical cytogenetics*. The remarkable discovery that a minute segment of the chromosome, the so-called band, is the cause of human disease has further enhanced our understanding of biologic processes. Many chromosomes previously thought to be abnormal have proven to be normal heteromorphic variants containing fractions of sedimented DNA (Verma and Dosik, 1987; Verma, 1988).

The major recognition of the role of chromosomes in clinical medicine was established in 1959 when Lejeune described the extra G-group chromosome in Down syndrome. During the last 3 decades, several dozen chromosomal syndromes have been described (Yunis, 1977; deGrouchy and Turleau, 1984). It is well known that the severity of birth defects depends on the size of the genetic material duplicated or deleted in humans. Some chromosomal abnormalities are lethal. More than 50% of abortuses contain some type of chromosomal damage. In the following pages we shall discuss the principal reasons for chromosomal analysis.

9.1 INDICATIONS FOR REFERRAL

Precise clinical diagnosis of individuals based on chromosomal abnormalities alone is difficult. Unfortunately, the clinical manifestation of features often shows great variation, depending on the type and size of chromosomes involved. Nevertheless, karyotype-phenotype correlation for complete trisomies and monosomies has been established for a number of chromosomes. In recent years, with a battery of banding techniques, several dozen chromosomal syndromes have been characterized. A detailed description of well-established syndromes is fully covered in various books. We need not know the specific syndromes associated with specific chromosomes; rather we should be aware of the features for which chromosomal analysis is indicated. A concise summary of some of these indications is as follows (Bocian and Mohandas, 1978):

1. To rule out the classical chromosomal syndromes.
2. To determine individuals with multiple congenital anomalies.
3. In parents of siblings with chromosomal abnormalities.

4. In all children of individuals with balanced or structural anomalies.
5. In couples with histories of two or more fetal losses.
6. In abortuses and malformed stillborns.
7. In couples with infertility problems.
8. In individuals with ambiguous genitalia.
9. In females with amenorrhea.
10. In pubertal failure in males and females.
11. In individuals with mental retardation.
12. In individuals with hematologic disorders.
13. For chromosomal analysis of amniotic fluids from mothers of advanced age or a history of genetic problems.
14. In individuals with chromosomal instability syndromes.
15. In individuals exposed to carcinogens, radiation, and the like.

9.2 CLASSICAL CHROMOSOMAL SYNDROMES

The era of chromosomal syndromes emerged in 1959 when Lejeune demonstrated the presence of an additional chromosome 21 in Down syndrome. Today several dozen syndromes are well defined and can be recognized clinically even before chromosomal analysis. Establishing a genotype-phenotype relationship for the presence or absence of small bands in humans has become an intellectual exercise, because these "syndromes" may exhibit considerable phenotypic variation. Although tremendous strides have been made towards understanding the chromosomal syndromes, many avenues in clinical cytogenetics remain to be explored. For example, numerous bands of the human genome overlap in their morphology, thus masking correct identification of additional genetic material. The mechanism(s) that produce aberrant chromosomes are still intriguing. However, through the use of molecular and cytogenetic markers, progress is being made towards establishing the identities of the extra chromosomes.

Parental age is a major factor in producing aneuploidy in humans (Vig and Sandberg, 1987). Nevertheless, genes predisposing to nondisjunction cannot be ignored. Radiation, viruses, and teratogenic agents have also been postulated as additional causes of aneuploidy. The most common chromosomal abnormalities in the human population are summarized in Table 9.2.1.

Table 9.2.1. Incidence of the Most Common Chromosomal
Abnormalities in Newborns*
(Thompson and Thompson, 1986)

Type	Incidence
Autosomal anomaly	
Trisomy 13	1/20,000
Trisomy 18	1/8,000
Trisomy 21	1/800
Other trisomy	1/50,000
Sex chromosomal anomaly	
XXYY; XXY; XXX	1/1000
XO	1/10,000
Translocation	
Balanced	1/500
Unbalanced	1/2,000
Total chromosomal abnormalities	1/200

*Reproduced with permission from W.B. Saunders Co., Philadelphia.

9.3 CYTOGENETICS OF SPONTANEOUS ABORTION

Numerous factors are associated with pregnancy fetal loss. In the early 1960s parental chromosomes in couples with habitual abortion were usually not studied until other etiologic factors had been excluded. Nevertheless, recent advances in cytogenetic techniques have led us to believe that couples with a history of more than two spontaneous first trimester abortions may carry balanced chromosomal rearrangements or structural chromosomal abnormalities, leading to duplications or deletions in the fetus (Venkatraj and Verma, 1987). Approximately 15% of all recognized pregnancies end in spontaneous abortion. The incidence of chromosomal abnormalities in these spontaneous abortions can be as high as 50%. The type of chromosomal abnormalities found in abortuses are summarized in Table 9.3.1. Monosomy for the X chromosome is the most common anomaly, whereas trisomy 16 is the second most frequent problem found in human abortuses. Triploidy and tetraploidy are also common; nevertheless, triploid fetuses rarely survive.

Table 9.3.1. Relative Frequencies of Chromosomal
Abnormalities in Abortuses*
(Thompson and Thompson, 1986)

Type	Frequency (%)
Trisomies	52
Trisomy	
14	3.7
15	4.2
16	16.4
18	3.0
21	4.7
22	5.7
Other	14.3
45,X	18.0
Triploid	17.0
Tetraploid	6.0
Unbalanced translocations	3.0
Other	4.0

*Reproduced with permission from W.B. Saunders Co., Philadelphia.

9.4 CHROMOSOMAL ANOMALIES IN PRENATAL DIAGNOSIS

The various reasons for referral for cytogenetic analysis have previously been summarized. However, advanced maternal age is still the major reason for cytogenetic analysis. Individuals with affected fetuses can be identified through amniocentesis, thus providing the option for selective abortion. It is most important that accurate genetic counseling be provided before amniocentesis. Detailed description of the implications of this test are given in many recent books (Milunsky, 1986; Nyhan and Sakati, 1987). The most frequent anomaly associated with maternal age (trisomy 21), abstracted from Thompson and Thompson (1986), is presented in Table 9.4.1.

9.5 CHROMOSOME ANOMALIES IN MENTAL RETARDATION

Chromosomal abnormalities are gross defects that result in major genetic imbalances. The metamorphosis of humans continues to remain a mystery, but development depends on the quality and quantity of one's genes. Individuals with mental retardation generally have a higher incidence of chromosomal abnormalities. There are at least 25 percent more retarded males than females in all but the most severe forms of mental retardation (Lubs, 1983), which

Table 9.4.1. Risk of Down Syndrome in Fetuses
at Amniocentesis and in Live Births*
(Thompson and Thompson, 1986)

Maternal Age (yr)	Frequency of Down's Syndrome	
	Fetuses	Live Births
–19	—	1/1550
20–24	—	1/1550
25–29	—	1/1050
30–34	—	1/700
35	1/350	1/350
36	1/260	1/300
37	1/200	1/225
38	1/160	1/175
39	1/125	1/150
40	1/70	1/100
41	1/35	1/85
42	1/30	1/65
43	1/20	1/50
44	1/13	1/40
45	1/25	1/25

*Reproduced with permission from W.B. Saunders Co., Philadelphia.

are X linked. The presence of the fragile-X has attracted the most attention among all the X-linked forms of mental retardation (Sutherland and Hecht, 1985; Hagerman and McBogg, 1983). X-linked mental retardation accounts for approximately 25 percent of retardation in males and 10 percent of mild retardation in females. Cytogenetic detection in these cases remains difficult because the marker X is found only in a minority of cells.

9.6 CHROMOSOMES OF HUMAN NEOPLASIA

The role of chromosomes in the transformation of a normal to a neoplastic cell was first postulated over 70 years ago. However, great strides in understanding the chromosomal basis of human neoplasias have been made only in the past two decades. The application of various banding techniques has resulted in the discovery of several dozen human neoplasms associated with specific chromosomal aberrations, thus implicating certain regions in the process of malignant transformation (Verma, 1988; Lebeau and Rowley, 1986; Yunis, 1986; Sandberg, 1986; Heim and Mitelman, 1987). Acquired abnormalities are consistent and shared by multiple cells, suggesting clonal origin. These findings have greatly stimulated interest in the cytogenetic aspects of cancer, establishing a definite correlation between specific chromosomal abnormalities and myeloproliferative anomalies (Table 9.6.1). A few examples were chosen to elucidate some principles of chromosomal change in neoplastic disorders, answering fundamental questions such as: are specific chromosomal aberrations related to etiologic factor(s) or to the type of target cells in human neoplasms?

Before the advent of banding techniques, only numerical and structural rearrangements were classified. Remarkable discoveries have resulted from application of the newer banding techniques. The use of early prometaphase chromosomes has permitted visualization of defects that previously remained undetected (Yunis et al, 1981).

Over 70 human neoplasms are known to have one of the more than 40 specific recurrent chromosomal abnormalities that can be either translocations, deletions, duplications, inversions, or trisomies. Translocations are seen in leukemias and non-Hodgkin's lymphoma, whereas deletions are found more frequently in myelodysplasia and solid tumors. Consistent chromosomal defects are an important parameter used in the prognosis of a specific neoplasm; however, some chromosomal abnormalities are shared among related disorders (Table

Table 9.6.1a. Neoplasms with a Known Consistent
Chromosomal Defect*

Diseases	Chromosomal Defects
Chronic myelogenous leukemia	t(9;22)(q34.1;q11.21)
Acute nonlymphocytic leukemia	
M1,M2	inv(3)(q21q25–27)
M1,M2,M4,M5,M6	−5 or del(5)(q13q31)+
M1,M2	t(6;9)(p22.2;q34)
M1,M2,M4,M5,M6	−7 or del(7)(q31.2q36)+
M1,M2,M4,M5,M6	+8
M2	t(8;21)(q22.1;q22.3)
M2,M4,M5a	t(9;11)(p22;q23)+
M1,M2	t(9;22)(q34.1;q11.21)
M3	t(15;17)(q22;q11.2?)
M2,M4,M5b	inv(16)(p13.1q22.1)
Myelodysplasia-preleukemia	
RA,RAEB-CMML	t(1;3)(p36;q21)
RA,RA-S,RAEB	t(2;11)(p11;q23)
RA	del(5)(q13q31)+
RA,RA-S,RAEB,RAEB-T,CMML	−7 or del(7)(q31q36)+
RAEB	del(7)(p11.2p22)
RA,RAEB	+8
RAEB,RAEB-T	del(9)(q13q22)
RA-S,RAEB	del(20)(q12q13)
Chronic lymphocytic leukemia	
B cell	t(11;14)(q13;q32.3)
B cell	+12
T cell	inv(14)(q11.2q32.3) or
	t(14;14)(q11.2;q32.3)
Acute lymphocytic leukemia	
L1,L2, pre-B cell	t(1;19)(q21–23;p13?)
L1,L2, null	t(4;11)(q21;q23)
L2, null	del(6)(q21q25)+
L3, B cell	t(8;14)(q24.1;q32.3)
L1,L2, null	t(9;22)(q34.1;q11.21)
L1,L2, T cell	t(11;14)(p13–14.1;q11.2–13)

*Reproduced with permission from Yunis (1986).

9.6.2). One of the most common chromosomal abnormalities is the translocation t(9;22) found primarily in patients with chronic myelogenous leukemia but also in those with acute lymphocytic leukemia and acute nonlymphocytic leukemia. Recent reviews on this subject are well covered by Verma (1988); therefore, only a brief and comprehensive tabulation is provided here.

It recently has become evident that many genetic factors play important roles in the etiology of pediatric malignancies (Arthur, 1986). Childhood cancer can be grouped into two classes: hereditary and nonhereditary. Among hereditary malignancies, retinoblastoma and Wilm's tumor are well-investigated diseases. Nonhereditary pediatric malignancies include: chronic myelogenous leukemia, acute nonlymphocytic leukemia, and acute lymphoblastic leukemia. Most, if not all, pediatric neoplasms have chromosomal abnormalities. Certain individuals are predisposed to develop neoplasms and are referred to as cancer prone. The term "chromosome breakage" or "instability syndrome" has been coined to describe these individuals (Ray and German, 1983). Faconi's anemia, ataxia telangiectasia, and Bloom syndrome are the most common disorders. Nonetheless, xeroderma pigmentosum, Werner's syndrome, Kostmann's agranulocytosis, and anemia can also be included in this group.

Many individuals have congenital anomalies that are well correlated with chromosomal

Table 9.6.1b. Neoplasma with a Known Consistent Chromosomal Defect*

Diseases	Chromosomal Defects
Non-Hodgkin's lymphoma	
Diffuse large noncleaved cell	del(6)(q21q25)+
Burkitt, non-Burkitt	t(8;14)(q24.1;q32.3)
Small cell, immunoblastic	
Small lymphocytic, B cell	del(11)(q14.2q23.3)
Small lymphocytic, B cell	t(11;14)(q13;q32.3)
Small lymphocytic, B cell	+12
Diffuse mixed small and large T cell	t(12;14)(q13;q32.3)
Small lymphocytic T cell	inv(14)(q11.2q32.3) or
Sezary syndrome, mycosis fungoides	t(14;14)(q11.2;q32.3)
Follicular small cleaved follicular cell and follicular mixed small and large cell	t(14;18)(q32.3;q21.3)
Carcinoma	
Melanoma	del(1)(p21–22p36)+
Neuroblastoma	del(1)(p31.2p36)
Small cell lung carcinoma	del(3)(p14p23)+
Renal cell carcinoma, familial	t(3;8)(p14.2;q24.1)
Renal cell carcinoma	t(5;14)(q13;q22)del(14)(q22q32)
Ovarian papillary cystadenocarcinoma	t(6;14)(q21;q24)
Carcinoma of prostate	del(10)(q23–24q36)
Ewing's sarcoma, neuroepithelioma, Askin's tumor	t(11;22)(q23;q11.23)
Constitutional Wilms tumor, Wilms tumor	del(11)(p13)
Seminoma, teratoma	i(12p)
Constitutional retinoblastoma, retinoblastoma, osteosarcoma	del(13)(q14.1)
Multiple endocrine carcinoma	del(20)(p12.2)
Benign solid tumor	
Mixed parotid gland tumor	t(3;8)(p21;q12)
Meningioma	−22

*Reproduced with permission from Yunis (1986).

defects. These individuals are at a higher risk of cancer than are those who are chromosomally normal. For example, individuals with the Down syndrome have an 11-fold increased risk of leukemia as compared with that of the normal population. Patients with aneuploidy for sex chromosomes are more prone to cancer than is the general population. There are occasional reports of neoplasm in patients with structural and numerical mosaicism. In retrospect, patients with skin cancer have a significantly increased rate of chromosome type aberrations (Mitelman, 1985).

9.7 CHROMOSOMAL BASIS OF ONCOGENESIS

In recent years astonishing progress has been made towards understanding the chromosomal basis of oncogenesis. The status of chromosomal localization of the cellular homology of retroviral oncogenes (c-onc) was recently reviewed extensively. The breakpoints of many consistent rearrangements that characterize individual cancers involve specific bands to which different c-onc genes have been assigned (Table 9.7.1). The role of oncogenes with respect to chromosomal aberrations has been the major factor in their localization (Verma, 1986). Highly significant data pertaining to translocations and activation of oncogenes are rapidly accumulating (Melchers and Potter, 1986). In many, if not all, cancers, both tumorigenesis and tumor progression are multifactorial events. Documentation of this hypothesis has evolved through the study of chromosomal deletions in retinoblastoma, neuroblastoma, and Wilms' tumors. There is now ample reason for excitement as conflicting evidence of specific genes in causing tumors is unfolding (Bishop, 1983). The chromosomal events that occur in

Table 9.6.2. Neoplasms with a Shared Single Recurrent Chromosomal Defect*

Chromosomal Defects	Diseases
del 1p	Neuroblastoma Melanoma
t(1;3)(p36;q21)	Myelodysplasia Acute nonlymphocytic leukemia, M4
inv(3)(q21q27)	Myelodysplasia Acute nonlymphocytic leukemia, M1, M2
t(4;11)(q21;q23)	Acute lymphocytic leukemia, L1, L2 Acyte myelomonocytic leukemia
del(5)(q13q31)	Myelodysplasia Acute nonlymphocytic leukemia, M1, M2, M4, M5, M6
del(6)(q21q25)	Diffuse large cell lymphoma Acute lymphocytic leukemia, L2
t(6;9)(p21.2;q34)	Acute nonlymphocytic leukemia, M1, M2
del(7)(q31.2q36)	Myelodysplasia Acute nonlymphocytic leukemia, M1, M2, M4, M5, M6
+8	Myelodysplasia Acute nonlymphocytic leukemia, M1, M2, M4, M5, M6
t(8;14)(q24.1;q32.3)	Burkitt's lymphoma Acute lymphocytic leukemia, L3 Small noncleaved non-Burkitt's lymphoma Immunoblastic lymphoma, B cell
t(8;21)(q22.1;q22.3)	Acute myelogenous leukemia, M2, M4
t(9;11)(p22;q23.3)	Acute monocytic leukemia, M5 Acute myelomonocytic leukemia, M4 Acute nonlymphocytic leukemia, M2
t(9;22)(q34.1;q11.21)	Chronic myelogenous leukemia Acute myelogenous leukemia, M1, M2 Acute lymphocytic leukemia, L1, L2
t(11;14)(q13.3;q32.3)	Chronic lymphocytic leukemia, B cell Small cell lymphocytic lymphoma, B cell Diffuse large cell lymphoma, B cell
t(11;22)(q23;q11.2)	Ewing's sarcoma Neuroepithelioma Askin's tumor
+12	Chronic lymphocytic leukemia, B cell Small cell lymphocytic lymphoma, B cell
iso12p	Seminoma Teratoma
del 13(q14)	Retinoblastoma Constitutional retinoblastoma Osteosarcoma
inv(14q11.2q32.3) or t(14;14)(q11.2;q32.3)	Chronic lymphocytic leukemia, T cell Small lymphocytic lymphoma, T cell Sezary syndrome Mycosis fungoides
t(14;18)(q32.3;q21.3)	Follicular small cleaved cell lymphoma Follicular mixed cell lymphoma Follicular large cell lymphoma
inv(16)(p13.1q22.1)	Acute monocytic leukemia, M5b Acute myelomonocytic leukemia, M4 Acute nonlymphocytic leukemia, M2

*Reproduced with permission from Yunis (1986).

Table 9.7.1. Localization of Proto-Oncogenes in Human Chromosomes*

Chromosomes	Localization	Oncogene
1	p13	N-ras
	p32	L-myc & B-lym-1
	p32 → p35	lck
	p34 → p36	fgr
	p36	src
	q22 → q24	ski
2	p13 → cen	rel
	p23.2 → q24	N-myc
3	p25	raf-1
4	q11 → q22	kit
	?	raf-2
5	p13 → p14	M-lvi-2
	q33 → q34	csf-1R
	q34	fms
6	p21	pim
	p23 → q12	K-ras-1
	q21	syr
	q22	ros
	q22 → q23	myb
	?	yes-2
7	p12 → p13	egf-R
	p14 → q21	A-raf-2
	p12 → p14	erb-B
	p15 → p22	ral
	pter → q22	pks-2
	q21 → q31	met
8	q13 → qter	lyn
	q22.1 → q22.3	mos
	q24	pvt-1
	q24.1	myc
9	q34.1	abl
11	p15.5	H-ras-1
	q13	bcl-1,int-2 & sea
	q23 → 24	est-1
12	p12.1	K-ras-2
	q12 → 13	int-1
	q13 → 14.3	gli
	q24.2	K-ras-2
13	q12	flt
14	q21 → q31	fos
	q32	akt-1
15	q26.1	fes
16	?	fos
17	q11 → q21.3	A-erb-1
	q21 → 22	nb1
	?	neu
18	q21.3	yes-1 & bcl-2
20	q11 → 12	hck
	q13.1	src-1
21	q22.3	ets-2
22	q12.3 → q13.1	sis & P-dgf-B
X	p21 → q11	A-raf-1
	pter → q28	H-ras-2
	pter → q22	pks-1
	q27	mcf-2

*Adapted from Verma (1986) and Bloomfield et al (1987).

Table 9.8.1. Chromosomal Localization of Selected Diseases Using Molecular Techniques

Chromosomes	Diseases	Chromosomes	Diseases
1	Charcot-Marie Tooth Disease (Type 1) Aucher's disease	X	Aarskog syndrome
			Adrenoleukodystrophy
2	Aniridia-1		Agammaglobulinemia, X-linked
3	GM$_1$ Gangliosidosis GM$_2$ Gangliosidosis Pseudocholinesterase deficiency		Albinism-deafness syndrome
			Choroideremia
			Chronic granulomatous disease
4	Huntington's disease		Cleft palate, X-linked
5	Polyosis coli, multiple Sandoff disease (Type 2) schizophrenia		Diabetes insipidus, nephrogenic
			Ectodermal dysplasia, hypohidrotic
6	Adrenogenital 21-hydroxylase deficiency, Spinocerebellar ataxia, Hemochromatosis		Hemophilia A
			Hemophilia B
7	Cystic fibrosis		Lesch-Nyhan syndrome
9	Tuberous sclerosis Friedreich's ataxia		Manic-depressive illness (recessive)
10	Gyrate atrophy (ornithine aminotransferase, Multiple endocrine neoplasia		Mucopolysaccharidosis-Hunter
			Muscular dystrophy-Becker/ Duchenne's
11	Aniridia-2, β-Thalassemias Manic-depressive illness Sickle cell disease		Muscular dystrophy-Emery-Dreifuss
12	Phenylketonuria, von Willebrand's disease		Norrie's disease
13	Retinoblastoma, Wilson's disease		Oculocerebrorenal syndrome (Lowe's disease)
14	α_1-Antitrypsin deficiency Holt-Oram syndrome		Ornithine transcarbamylase deficiency
15	Tay-Sachs disease (type 1)		Pelizaeus-merzbacher disease
16	Polycystic Kidney disease, adult, α-Thalassemias		Retinitis pigmentosa, X-linked-
			Retinitis pigmentosa, X-linked-
17	Neurofibromatosis, peripheral		Retinoschisis
18	Amyloidotic polyneuropathy		Rickets, hypophosphatemic
19	Myotonic dystrophy		Testicular feminization, androgeninsensitive
21	Alzheimer's disease		
22	Meningioma		Wiskott-Aldrich syndrome

these three tumors may not be limited. Clearly these observations indicate the more widespread importance of these events in oncogenesis. The lack of an adequate mechanistic explanation for the developmental processes and their derangement should not stifle research in this fascinating area.

9.8   ANALYSIS OF GENOME WITH PROBES

The past two decades have revolutionized the field of molecular genetics because of the discoveries of newer recombinant DNA techniques (Childs et al, 1988; Davis, 1986; Galton, 1985). The application of DNA probes has resulted in the localization of various restriction fragment length polymorphisms (RFLPs) in the human genome (Landegren et al, 1988; Pearson et al, 1987). Through the use of linkage data analysis, various RFLPs have been associated with several diseases. Consequently, chromosomal localization of genes responsible for various human diseases has been identified (Martin, 1987; White and Lalouel, 1988). Single-base pair substitution and repeated DNA sequences have provided the molecular basis for the detection of RFLPs. The likelihood of a linkage between probe and defective gene is expressed as a ratio of the probability of linkage, or a logarithm of the odds (LOD score). The purpose of this book is not to cover the theoretic aspects of gene localization, but to highlight the techniques that are used most commonly. For detailed descriptions, investigators are advised to consult highly specialized publications (Maniatis et al, 1982; Wu, 1979; Ott, 1985). Briefly, various diseases that have been mapped are summarized in Table 9.8.1. In the coming years, the precise location of various diseases on the human genome will increase greatly. Gene therapy of genetic defects is still in its infancy; however, early results in vitro and in animal models are promising (Rosenberg, 1985). Proteins of known therapeutic value, developed by commercial companies using recombinant DNA technology, hold enormous promise for therapeutic intervention of diseases in clinical medicine. Many of these products will one day enter the domain of primary care providers.

9.9 REFERENCES

Arthur DC: Genetics and cytogenetic of pediatric cancers. *Cancer* 1986;58:534–540.

Bishop JM: Cancer genes come of age. *Cell* 1983;32:1018–1020.

Bloomfield C D, Trent JM, Van den Berghe H: Report of the committee on structural chromosome changes in neoplasia. *Cytogenet Cell Genet* 1987;46:344–366.

Bocian M, Mohansas T: Recent cytogenetic advances and implications for pediatric practice in Kabach MM (ed): *The Pediatric Clinics of North America*, Philadelphia, W.B. Saunders, 1978, vol 25, pp 517–538.

Childs B, Holtzman NA, Kazazian HH, Valle DL (eds): *Molecular Genetics in Medicine*. New York, Elsevier Science Publishing Co., 1988, vol 7, pp 18–31.

Davies KE (ed): *Human Genetic Diseases*. Oxford, UK, IRL Press, 1986.

DeGrouchy J, Turleau C: *Clinical Atlas of Human Chromosomes*. New York, John Wiley & Sons, 1984.

Davis LG, Dibner MD, Battey JF: *Basic Methods in Molecular Biology*. New York, Elsevier Science Pub. Co., 1986.

Galton DJ: *Molecular Genetics of Common Metabolic Diseases*. John Wiley & Sons, New York, 1985.

Hagerman RJ, McBogg PM (eds): *The Fragile X Syndrome*. Dillon, Colorado Spectra Publishing Co., 1983.

Heim S, Mitelman F: *Cancer Cytogenetics*. New York, Alan R. Liss Inc., 1987.

Landegren V, Kaiser R, Caskey CT, Hood L: DNA diagnostics—Molecular technique and automation. *Science* 1988;242:229–237.

LeBeau MM, Rowley JD: Chromosomal abnormalities in leukemia and lymphoma: Clinical and biological significance, in Harris H, Hirschhorn K (eds): *Advances in Human Genetics*. New York, Plenum Publishing Co., vol 15, pp 1–54.

Lejeune J: Le monoglisme: Premier example d'aberration autosomique humaine. *Ann Genet Sem Hop* 1959;1:41–49.

Lubs HA: X-linked mental retardation and the marker X, in Emery AEH, Rimoin DL (eds): *Principles and Practice of Medical Genetics*. London, Churchill Livingstone, 1983, vol 1, pp 216–223.

Maniatis T, Fritsh EF, Sambrook J: *Molecular Cloning: A Laboratory Manual*. New York, Cold Springs Harbor Laboratory, 1982.

Martin JB: Molecular genetics: Application to the clinical neurosciences. *Science* 1987;238:765–772.

Melchers F, Potter M (eds): *Mechanisms in B-Cell Neoplasia*. New York, Springer-Verlag, 1986.

Milunsky A (ed): *Genetic Disorder and the Fetus: Diagnosis, Prevention and Treatment*. New York, Plenum Press, 1986, pp 115–172.

Mitelman F: *Catalog of Chromosome Aberrations in Cancer*. New York, Alan R. Liss Inc., 1985.

Nyhan WL, Sakati NA: *Diagnostic Recognition of Genetic Disease*. Philadelphia, Lea and Febiger, 1987.

Ott J: *Analysis of Human Genetic Linkage*. Baltimore, Johns Hopkins University Press, 1985.

Pearson PL, Kidd KK, Willard HF: Human gene mapping by recombinant DNA techniques. *Cytogenet Cell Genet* 1987;46:390–566.

Ray JH, German J: The cytogenetics of the chromosome—breakage syndrome, in Germun J (ed): *Chromosome Mutation and Neoplasia*. New York, Alan R. Liss Inc., 1983, pp 97–134.

Rosenberg RN: *Neurogenetics: Principles and Practice*. New York, Raven Press, 1985.

Sandberg AA: The chromosomes in human leukemia. *Sem Hematol* 1986;23:201–217.

Sutherland GR, Hecht F: *Fragile Sites on Human Chromosomes*. New York, Oxford University Press, 1985.

Thompson JS, Thompson MW: *Genetics in Medicine*. Philadelphia, W.B. Saunders, 1986.

Venkatraj VS, Verma RS: Chromosomal breakpoints in aborters: A relationship with heritable fragile site. *Gynecol Obstet Invest* 1987;24:241–249.

Verma RS: Chromosomal abnormalities in neoplastic diseases, in Rosenthal CJ (ed): *Manual of Clinical Oncology*, New York, Elsevier Scientific Publication, 1988 (in press).

Verma RS: Oncogenetics: A new emerging field of cancer. *Mol Gen Genet* 1986;205:385–389.

Verma RS (ed): *Heterochromatin: Molecular and Structural Aspects*. New York, Cambridge University Press, 1988.

Verma RS, Dosik H: Structural organization of heterochromatin in the human genome, in G.H. Bourne (ed): *Cytology and Cell Physiology*. New York, Academic Press, 1987, 4th ed., pp 685–705.

Vig BK, Sandberg AA (eds): *Aneuploidy*. New York, Alan R. Liss, 1987.

White R, Lalouel JM: Chromosome mapping with DNA markers. *Sci Am* 1988;258:40–48.

Wu R (ed): Recombinant DNA, in *Methods in Enzymology*. New York, Academic Press, vol 68, 1979.

Yunis JJ: *New Chromosomal Syndrome*. New York, Academic Press, 1977.

Yunis JJ: Chromosomal rearrangements, gene, and fragile sites in cancer: Clinical and biological implications, in Devita VT, Hellman S, Rosenberg SA (eds): *Important Advances in Oncology*. Philadelphia, J.B. Lippincott Co., 1986, pp 93–128.

Yunis JJ, Bloomfield CD, Ensrud K: All patients with acute non-lymphocytic leukemia may have a chromosomal defect. *N Engl J Med* 1981;305:135–139.

Index